BIODEGRADABLE WASTE AND MANAGEMENT

Edited by
Alejandro White

Arcler Press

Biodegradable Waste and Management

Edited by Alejandro White

Edition 2018

ArclerPress

Published by
Arcler Press LLC
708 3rd Avenue, 6th Floor
New York
NY 10017
United States of America
www.arclerpress.com

ISBN: 978-1-68094-615-4 (Hardback)

Printed in United States of America

Library of Congress Control Number: 2017933214

About the Editor

Alejandro White

Alejandro is an Environmental Engineering BSc. from Universidad de los Andes, Colombia. His background and experience include areas like air quality and pollution, purification and waste water treatment and solid waste management. His interests span from sustainable architecture to renewable energies. Currently he is settled in the Philippines working for an international non-governmental organization after his passage from the private sector working in Coca Cola FEMSA.

Table of Contents

4 COMPOST MANAGEMENT

5 AEROBIC DIGESTION

6 ANAEROBIC DIGESTION

7 BIOCONVERSION SYSTEMS

8 WASTE COLLECTION AND MANAGEMENT

9 MECHANICAL BIOLOGICAL TREATMENT

List of Abbreviations

ACP	`Acyl carrier protein
ATP	Adenosine triphosphate
AADF	Annual average design flow
ARGUS	Aerobic granules upgrade system)
ARB	Association of Bogotá
BOD	Biochemical Oxygen Demand
BNR	Biological nutrient removal
BPA	Bisphenol A
COD	Chemical oxygen demand
DOE	Department of Energy's
FAMEs	Fatty acid methyl (or ethyl) esters
GSHP	Ground source heat pump
ISO	International Standards Organization
MBT	Mechanical biological treatment
MBBR	Moving bed biofilm reactor
MSW	Municipal solid waste,
PBR	Photobioreactor
PSA	Pressure swing adsorption
PPP	Public-Private Partnership
RAS	Return activated sludge
SBBGR	Sequencing batch biofilter granular reactor
SBR	Sequencing batch reactors
STP	Sewage treatment plants

TSS	Total suspended solids
TPWES	Transportable Plasma Waste to Energy System
UV	Ultraviolet
UDDT	Urine-diverting dry toilets
VFAs	Volatile fatty acids

Preface

Biodegradable waste is a type of waste, typically originating from plant or animal sources, which may be degraded by other living organisms. Waste that cannot be broken down by other living organisms are called non-biodegradable. Biodegradable waste can be commonly found in municipal solid waste (sometimes called biodegradable municipal waste, or BMW) as green waste, food waste, paper waste, and biodegradable plastics. Other biodegradable wastes include human waste, manure, sewage, slaughterhouse waste. In the absence of oxygen much of this waste will decay to methane by anaerobic digestion. The main environmental threat from biodegradable waste is the production of methane in landfills. Unquestionably, landfilling is the worst waste management option for bio-waste. However, for the management of biodegradable waste diverted from landfills, there seems to be several environmentally favorable options. While the waste management hierarchy also applies to the management of bio-waste, in specific cases it may be justified to depart from it as the environmental balance of the various options available for the management of this waste depends on a number of local factors, inter alia collection systems, waste composition and quality, climatic conditions, the potential of use of various waste-derived products such as electricity, heat, methane-rich gas or compost. Therefore, national strategies for the management of this waste should be determined in a transparent manner and be based on a structured and comprehensive approach such as Life Cycle Thinking (LCT). Biodegradable waste can often be used for composting which turns the waste into humus. It can also be used as a resource for heat, electricity and fuel in future, by using anaerobic digestion. This produces additional Biogas and still delivers usable nutrients that can be implemented to the soil. Especially animal-based kitchen scraps are best disposed this way. The leftover human food can also be used as fodder to flightless birds, goats, pigs, sheep, or alternatively can be fed to birds, and other wildlife.

Content Coverage

Chapter One is intended to present an introduction of biodegradable waste that can be used for composting or a resource for heat, electricity and fuel by means of incineration or anaerobic digestion.

Chapter Two provides the information about biodegradable polymers, which have a long history and since many are natural products, the precise timeline of their discovery and use cannot be accurately traced.

Chapter Three explores about the biofuels that is produced through contemporary biological processes, such as agriculture and anaerobic digestion, rather than a fuel produced by geological processes such as those involved in the formation of fossil fuels, such as coal and petroleum, from prehistoric biological matter.

Chapter Four outlines the principles of composting of waste, an aerobic method of decomposing solid wastes. The process involves decomposition of organic waste into humus known as compost which is a good fertilizer for plants.

Chapter Five elaborates the process of aerobic digestion typically used in an activated sludge treatment plant. Waste activated sludge and primary sludge are combined, where appropriate, and passed to a thickener where the solids content is increased.

Chapter Six presents the anaerobic digestion, which is widely used as a source of renewable energy. The process produces a biogas, consisting of methane, carbon dioxide and traces of other 'contaminant' gases. This biogas can be used directly as fuel, in combined heat and power gas engines or upgraded to natural gas-quality biomethane. The nutrient-rich digestate also produced can be used as fertilizer.

Chapter Seven focuses on bioconversion systems. Bioconversion is the conversion of organic materials, such as plant or animal waste, into usable products or energy sources by biological processes or agents, such as certain microorganisms.

Chapter Eight emphasizes on the process of waste collection. It is the transfer of solid waste from the point of use and disposal to the point of treatment or landfill. Waste collection also includes the curbside collection of recyclable materials that technically are not waste, as part of a municipal landfill diversion program.

Chapter Nine focuses on mechanical biological treatment (MBT) system, a type of waste processing facility that combines a sorting facility with a form of biological treatment such as composting or anaerobic digestion. MBT plants are designed to process mixed household waste as well as commercial and industrial wastes.

CHAPTER 1

Biodegradable Waste: General Overview

Chapter Contents

- **Introduction**
- **Sources**
- **Uses of Biodegradable Waste**
- **Areas Relying on Organic Waste**
- **Climate Change Impacts**
- **Biodegradation**
- **Sewage Treatment**

The capacity of a natural environment to chemically break down an object is referred to as *biodegradability*. Some materials break down into organic parts far faster than others, leading them to be referred to as *biodegradable*. Measuring the biodegradability of an object can determine how long it will last in its current form.

The process that causes biodegradability is essentially decomposition, or rotting. A buried or abandoned object is slowly taken apart by surrounding microorganisms, bacteria, and exposure to the elements. While observable breakdown is visible in some items within hours, other materials may remain essentially unchanged for centuries or even millennia. This leads to considerable concern about the amount of non-biodegradable trash created through human consumption, and has lead to a push to use more biodegradable products.

Typically, materials constructed from plant or animal material have a high rate of biodegradability. Paper, made mostly from tree pulp, will degrade fairly quickly in a natural environment, because the Earth naturally contains the microorganisms to break such material down. Materials that are primarily man-made, on the other hand, do not necessarily have naturally occurring mechanisms to break the material down.

INTRODUCTION

Biodegradable waste includes any organic matter in waste which can be broken down into carbon dioxide, water, methane or simple organic molecules by micro-organisms and other living things using composting, aerobic digestion, anaerobic digestion or similar processes. In waste management, it also includes some inorganic materials which can be decomposed by bacteria. Such materials include gypsum and its products such as plasterboard and other simple organic sulfates which can be decomposed to yield hydrogen sulfide in anaerobic land-fill conditions.

In domestic waste collection, the scope of biodegradable waste may be narrowed to include only those degradable wastes capable of being handled in the local waste handling facilities.

SOURCES

Biodegradable waste can be commonly found in municipal solid waste (sometimes called biodegradable municipal waste, or BMW) as green waste, food waste, paper waste, and biodegradable plastics. Other biodegradable wastes include human waste, manure, sewage, sewage sludge and slaughterhouse waste. In the absence of oxygen, much of this waste will decay to methane by anaerobic digestion.

In many parts of the developed world, biodegradable waste is separated from the rest of the waste stream, either by separate kerb-side collection or by waste sorting after collection. At the point of collection such waste is often referred to as *Green waste*. Removing such waste from the rest of the waste stream substantially reduces waste volumes for disposal and also allows biodegradable waste to be composted where composting facilities exist.

USES OF BIODEGRADABLE WASTE

Biodegradable waste can be used for composting or a resource for heat, electricity and fuel by means of incineration or anaerobic digestion. Swiss *Kompogas* and the Danish *AIKAN* process are examples of anaerobic digestion of biodegradable waste. While incineration can recover the most energy, anaerobic digestion plants retain the nutrients and compost for the soil and still recover some of the contained energy in the form of biogas. Kompogas produced 27 million Kwh of electricity and biogas in 2009. The oldest of the company's own lorries has achieved 1,000,000 kilometers driven with biogas from household waste in the last 15 years.

AREAS RELYING ON ORGANIC WASTE

Featured in an edition of *The Economist* that predicted events in 2014, it was revealed that Massachusetts creates roughly 1.4 million tons of organic waste every year. Massachusetts, along with Connecticut and Vermont, are also going to enact laws to divert food waste from landfills.

In small and densely populated states, landfill capacity is limited so disposal costs are higher ($60–90 per ton in MA compared to national average of $45). Decomposing food waste generates methane, a notorious greenhouse gas. However, this biogas can be captured and turned into energy through anaerobic digestion, and then sold into the electricity grid.

Anaerobic digestion grew in Europe, but is starting to develop in America. Massachusetts is increasing its production of anaerobic digesters.

CLIMATE CHANGE IMPACTS

The main environmental threat from biodegradable waste is the production of methane and other greenhouse gases.

BIODEGRADATION

Yellow slime mold growing on a bin of wet paper.

Biodegradation is the disintegration of materials by bacteria, fungi, or other biological means. Although often conflated, biodegradable is distinct in meaning from compostable. While biodegradable simply means to be consumed by microorganisms, compostable makes the specific demand that the object break down under composting conditions. The term is often used in relation to ecology, waste management, biomedicine, and the natural environment (bioremediation) and is now commonly associated with environmentally friendly products that are capable of decomposing back into natural elements. Organic material can be degraded aerobically with oxygen, or anaerobically, without oxygen. Biosurfactant, an extracellular surfactant secreted by microorganisms, enhances the biodegradation process.

Biodegradable matter is generally organic material that serves as a nutrient for microorganisms. Microorganisms are so numerous and diverse that, a huge range of compounds are biodegraded, including hydrocarbons (e.g. oil), polychlorinated biphenyls

(PCBs), polyaromatic hydrocarbons (PAHs), pharmaceutical substances. Decomposition of biodegradable substances may include both biological and abiotic steps.

Factors Affecting Rate

In practice, almost all chemical compounds and materials are subject to biodegradation, the key is the relative rates of such processes - minutes, days, years, centuries... A number of factors determine the degradation rate of organic compounds. Salient factors include light, water and oxygen. Temperature is also important because chemical reactions proceed more quickly at higher temperatures. The degradation rate of many organic compounds is limited by their bioavailability. Compounds must be released into solution before organisms can degrade them.

Biodegradability can be measured in a number of ways. Respirometry tests can be used for aerobic microbes. First one places a solid waste sample in a container with microorganisms and soil, and then aerate the mixture. Over the course of several days, microorganisms digest the sample bit by bit and produce carbon dioxide – the resulting amount of CO_2 serves as an indicator of degradation. Biodegradability can also be measured by anaerobic microbes and the amount of methane or alloy that they are able to produce. In formal scientific literature, the process is termed bio-remediation.

Approximated time for compounds to biodegrade in a marine environment	
Product	**Time to Biodegrade**
Paper towel	2–4 weeks
Newspaper	6 weeks
Apple core	2 months
Cardboard box	2 months
Wax coated milk carton	3 months
Cotton gloves	1–5 months
Wool gloves	1 year
Plywood	1–3 years
Painted wooden sticks	13 years
Plastic bags	10–20 years
Tin cans	50 years
Disposable diapers	50–100 years
Plastic bottle	100 years
Aluminium cans	200 years
Glass bottles	Undetermined

Detergents

In advanced societies, laundry detergents are based on *linear* alkylbenzenesulfonates. Branched alkybenzenesulfonates (below right), used in former times, were abandoned because they biodegrade too slowly.

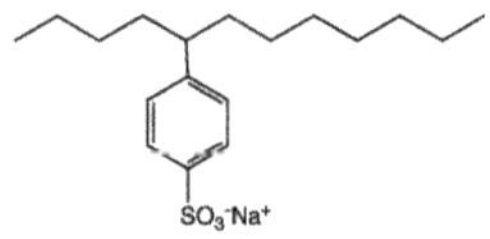

4-(5-Dodecyl) benzenesulfonate, a linear dodecylbenzenesulfonate

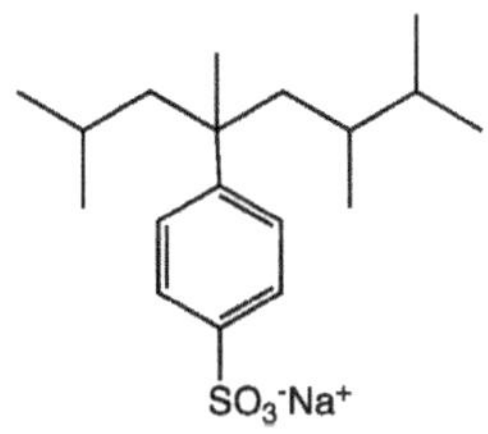

A branched dodecylbenzenesulfonate, which has been phased out in developed countries.

Plastics

Plastics biodegrade at highly variable rates. PVC-based plumbing is specifically selected for handing sewage because PVC biodegrades very slowly. Some packaging materials on the other hand are being developed that would degrade readily upon exposure to the environment. Illustrative synthetic polymers that are biodegrade quickly include polycaprolactone, others are polyesters and aromatic-aliphatic esters, due to their ester bonds being susceptible to attack by water. A prominent example is poly-3-hydroxybutyrate, the renewably derived polylactic acid, and the synthetic polycaprolactone. Others are the cellulose-based cellulose acetate and celluloid (cellulose nitrate).

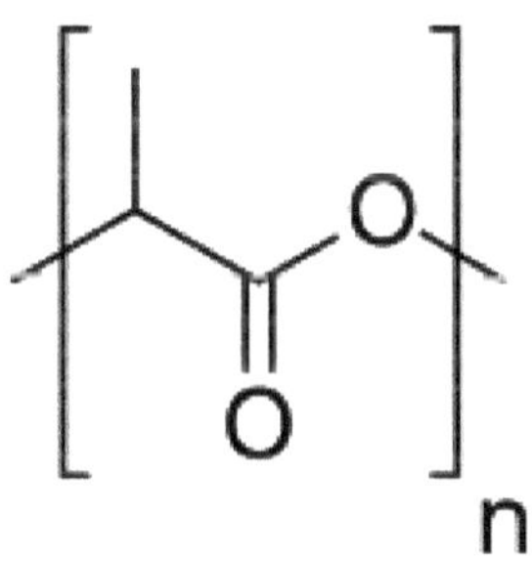

Polylactic acid is an example of a plastic that biodegrades quickly.

Under low oxygen conditions biodegradable plastics break down slower and with the production of methane, like other organic materials do. The breakdown process is accelerated in a dedicated compost heap. Starch-based plastics will degrade within two to four months in a home compost bin, while polylactic acid is largely undecomposed, requiring higher temperatures. Polycaprolactone and polycaprolactone-starch composites decompose slower, but the starch content accelerates decomposition by leaving behind a porous, high surface area polycaprolactone. Nevertheless, it takes many months. In 2016, a bacterium named Ideonella sakaiensis was found to biodegrade PET.

Many plastic producers have gone so far even to say that their plastics are compostable, typically listing corn starch as an ingredient. However, these claims are questionable because the plastics industry operates under its own definition of compostable:

that which is capable of undergoing biological decomposition in a compost site such that the material is not visually distinguishable and breaks down into carbon dioxide, water, inorganic compounds and biomass at a rate consistent with known compostable materials. (Ref: ASTM D 6002)

The term composting is often used informally to describe the biodegradation of packaging materials. Legal definitions exist for compostability, the process that leads to compost. Four criteria are offered by the European Union:

- Biodegradability, the conversion of >90% material material into CO2 and water by the action of micro-organisms within 6 months.
- Disintegrability, the fragmentation of 90% of the original mass to particles that then pass through a 2 mm sieve.
- Absence of toxic substances and other substances that impede composting.

Biodegradable Technology

In 1973 it was proven for the first time that polyester degrades when disposed in bioactive material such as soil. Polyesters are water resistant and can be melted and shaped into sheets, bottles, and other products, making certain plastics now available as a biodegradable product. Following, Polyhydroxylalkanoates (PHAs) were produced directly from renewable resources by microbes. They are approximately 95% cellular bacteria and can be manipulated by genetic strategies. The composition and biodegradability of PHAs can be regulated by blending it with other natural polymers. In the 1980s the company ICI Zenecca commercialized PHAs under the name Biopol. It was used for the production of shampoo bottles and other cosmetic products. Consumer response was unusual. Consumers were willing to pay more for this product because it was natural and biodegradable, which had not occurred before.

Now biodegradable technology is a highly developed market with applications in product packaging, production and medicine. Biodegradable technology is concerned with the manufacturing science of biodegradable materials. It imposes science-based mechanisms of plant genetics into the processes of today. Scientists and manufacturing corporations can help impact climate change by developing a use of plant genetics that would mimic some technologies. By looking to plants, such as biodegradable material harvested through photosynthesis, waste and toxins can be minimized.

Oxo-biodegradable technology, which has further developed biodegradable plastics, has also emerged. Oxo-biodegradation is defined by CEN (the European Standards Organisation) as degradation resulting from oxidative and cell-mediated phenomena, either simultaneously or successively." Whilst sometimes described as "oxo-fragmentable, and oxo-degradable this describes only the first or oxidative phase. These descriptions should not be used for material which degrades by the process of oxo-biodegradation defined by CEN, and the correct description is oxo-biodegradable.

By combining plastic products with very large polymer molecules, which contain only carbon and hydrogen, with oxygen in the air, the product is rendered capable of decomposing in anywhere from a week to one to two years. This reaction occurs even without prodegradant additives but at a very slow rate. That is why conventional plastics, when discarded, persist for a long time in the environment. Oxo-biodegradable formulations catalyze and accelerate the biodegradation process but it takes considerable skill and experience to balance the ingredients within the formulations so as to provide the product with a useful life for a set period, followed by degradation and biodegradation.

Biodegradable technology is especially utilized by the bio-medical community. Biodegradable polymers are classified into three groups: medical, ecological, and dual application, while in terms of origin they are divided into two groups: natural and synthetic. The Clean Technology Group is exploiting the use of supercritical carbon dioxide, which under high pressure at room temperature is a solvent that can use biodegradable plastics to make polymer drug coatings. The polymer (meaning a material composed of molecules with repeating structural units that form a long chain) is used to encapsulate a drug prior to injection in the body and is based on lactic acid, a compound normally produced in the body, and is thus able to be excreted naturally. The coating is designed for controlled release over a period of time, reducing the number of injections required and maximizing the therapeutic benefit. Professor Steve Howdle states that biodegradable polymers are particularly attractive for use in drug delivery, as once introduced into the body they require no retrieval or further manipulation and are degraded into soluble, non-toxic by-products. Different polymers degrade at different rates within the body and therefore polymer selection can be tailored to achieve desired release rates.

Other biomedical applications include the use of biodegradable, elastic shape-memory polymers. Biodegradable implant materials can now be used for minimally invasive surgical procedures through degradable thermoplastic polymers. These polymers are now able to change their shape with increase of temperature, causing shape memory capabilities as well as easily degradable sutures. As a result, implants can now fit through small incisions, doctors can easily perform complex deformations, and sutures and other material aides can naturally biodegrade after a completed surgery.

Etymology of "Biodegradable"

The first known use of the word in biological text was in 1961 when employed to describe the breakdown of material into the base components of carbon, hydrogen, and oxygen by microorganisms. Now biodegradable is commonly associated with environmentally friendly products that are part of the earth's innate cycle and capable of decomposing back into natural elements.

SEWAGE TREATMENT

Sewage treatment is the process of removing contaminants from wastewater, primarily from household sewage. It includes physical, chemical, and biological processes to remove these contaminants and produce environmentally safe treated wastewater (or treated effluent). A by-product of sewage treatment is usually a semi-solid waste or slurry, called sewage sludge that has to undergo further treatment before being suitable for disposal or land application.

Wastewater treatment plant in Massachusetts, United States

Sewage treatment may also be referred to as wastewater treatment, although the latter is a broader term which can also be applied to purely industrial wastewater. For most cities, the sewer system will also carry a proportion of industrial effluent to the sewage treatment plant which has usually received pretreatment at the factories themselves to reduce the pollutant load. If the sewer system is a combined sewer then it will also carry urban runoff (stormwater) to the sewage treatment plant.

Terminology

The term “sewage treatment plant” (or “sewage treatment works” in some countries) is nowadays often replaced with the term “wastewater treatment plant”.

Sewage can be treated close to where the sewage is created, which may be called a “decentralized” system or even an “on-site” system (in septic tanks, biofilters or aerobic treatment systems). Alternatively, sewage can be collected and transported by a network of pipes and pump stations to a municipal treatment plant. This is called a “centralized” system (see also sewerage and pipes and infrastructure),

Origins of Sewage

Sewage is generated by residential, institutional, commercial and industrial establishments. It includes household waste liquid from toilets, baths, showers, kitchens, and sinks draining into sewers. In many areas, sewage also includes liquid waste from industry and commerce. The separation and draining of household waste into greywater and blackwater is becoming more common in the developed world, with treated greywater being permitted to be used for watering plants or recycled for flushing toilets.

Sewage Mixing with Rainwater

Sewage may include stormwater runoff or urban runoff. Sewerage systems capable of handling storm water are known as combined sewer systems. This design was common when urban sewerage systems were first developed, in the late 19th and early 20th centuries.[:119] Combined sewers require much larger and more expensive treatment facilities than sanitary sewers. Heavy volumes of storm runoff may overwhelm the sewage treatment system, causing a spill or overflow. Sanitary sewers are typically much smaller than combined sewers, and they are not designed to transport stormwater. Backups of raw sewage can occur if excessive infiltration/inflow (dilution by stormwater and/or groundwater) is allowed into a sanitary sewer system. Communities that have urbanized in the mid-20th century or later generally have built separate systems for sewage (sanitary sewers) and stormwater, because precipitation causes widely varying flows, reducing sewage treatment plant efficiency.

As rainfall travels over roofs and the ground, it may pick up various contaminants including soil particles and other sediment, heavy metals, organic compounds, animal waste, and oil and grease. Some jurisdictions require stormwater to receive some level of treatment before being discharged directly into waterways. Examples of treatment processes used for stormwater include retention basins, wetlands, buried vaults with various kinds of media filters, and vortex separators (to remove coarse solids).

Industrial Effluent

In highly regulated developed countries, industrial effluent usually receives at least pretreatment if not full treatment at the factories themselves to reduce the pollutant load, before discharge to the sewer. This process is called industrial wastewater treatment. The same does not apply to many developing countries where industrial effluent is more likely to enter the sewer if it exists, or even the receiving water body, without pretreatment.

Industrial wastewater may contain pollutants which cannot be removed by conventional sewage treatment. Also, variable flow of industrial waste associated with production cycles may upset the population dynamics of biological treatment units, such as the activated sludge process.

Process Steps

Sewage collection and treatment is typically subject to local, state and federal regulations and standards. Treating wastewater has the aim to produce an effluent that will do as little harm as possible when discharged to the surrounding environment, thereby preventing pollution compared to releasing untreated wastewater into the environment.

Sewage treatment generally involves three stages, called primary, secondary and tertiary treatment.

- Primary treatment consists of temporarily holding the sewage in a quiescent basin where heavy solids can settle to the bottom while oil, grease and lighter solids float to the surface. The settled and floating materials are removed and the remaining liquid may be discharged or subjected to secondary treatment. Some sewage treatment plants that are connected to a combined sewer system have a bypass arrangement after the primary treatment unit. This means that during very heavy rainfall events, the secondary and tertiary treatment systems can be bypassed to protect them from hydraulic overloading, and the mixture of sewage and stormwater only receives primary treatment.
- Secondary treatment removes dissolved and suspended biological matter. Secondary treatment is typically performed by indigenous, water-borne micro-organisms in a managed habitat. Secondary treatment may require a separation process to remove the micro-organisms from the treated water prior to discharge or tertiary treatment.
- Tertiary treatment is sometimes defined as anything more than primary and secondary treatment in order to allow rejection into a highly sensitive or fragile ecosystem (estuaries, low-flow rivers, coral reefs,...). Treated water is sometimes disinfected chemically or physically (for example, by lagoons and microfiltration) prior to discharge into a stream, river, bay, lagoon or wetland, or it can be used for the irrigation of a golf course, green way or park. If it is sufficiently clean, it can also be used for groundwater recharge or agricultural purposes.

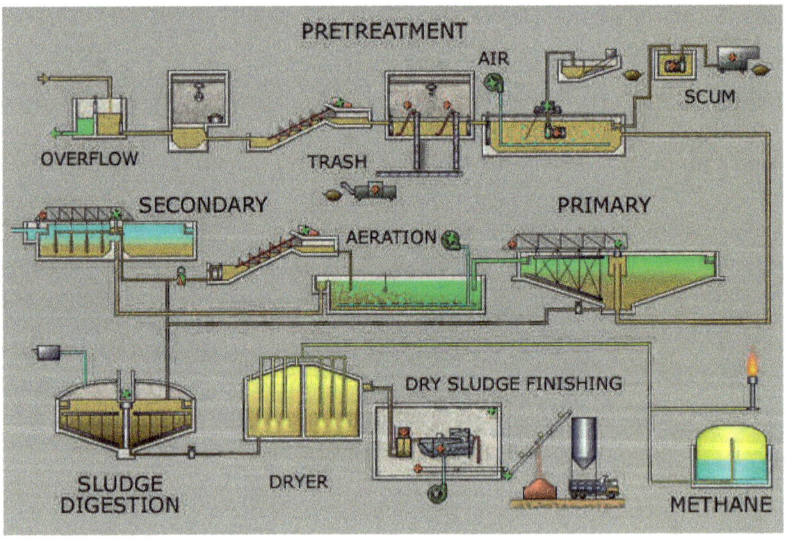

Simplified process flow diagram for a typical large-scale treatment plant.

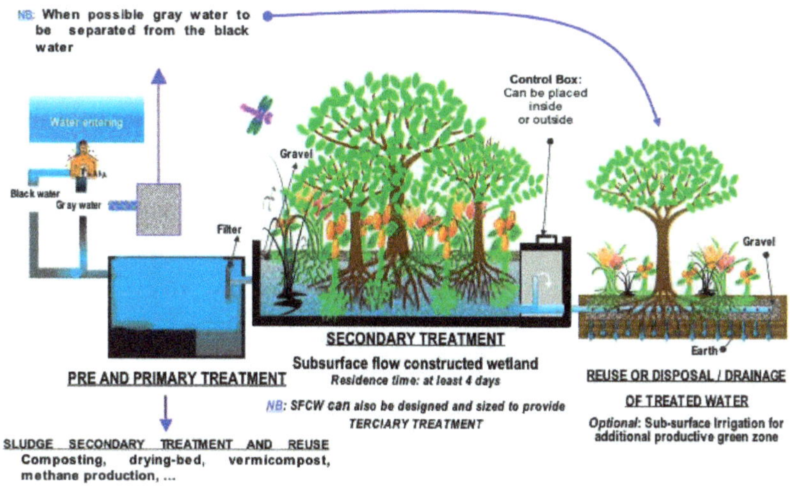

Process flow diagram for a typical treatment plant via subsurface flow constructed wetlands (SFCW)

Pretreatment

Pretreatment removes all materials that can be easily collected from the raw sewage before they damage or clog the pumps and sewage lines of primary treatment clarifiers. Objects commonly removed during pretreatment include trash, tree limbs, leaves, branches, and other large objects.

The influent in sewage water passes through a bar screen to remove all large objects like cans, rags, sticks, plastic packets etc. carried in the sewage stream. This is most commonly done with an automated mechanically raked bar screen in modern plants serving large populations, while in smaller or less modern plants, a manually cleaned screen may be used. The raking action of a mechanical bar screen is typically paced according to the accumulation on the bar screens and/or flow rate. The solids are collected and later disposed in a landfill, or incinerated. Bar screens or mesh screens of varying sizes may be used to optimize solids removal. If gross solids are not removed, they become entrained in pipes and moving parts of the treatment plant, and can cause substantial damage and inefficiency in the process.

Grit Removal

Pretreatment may include a sand or grit channel or chamber, where the velocity of the incoming sewage is adjusted to allow the settlement of sand, grit, stones, and broken glass. These particles are removed because they may damage pumps and other equipment. For small sanitary sewer systems, the grit chambers may not be necessary, but grit removal is desirable at larger plants. Grit chambers come in 3 types: horizontal grit chambers, aerated grit chambers and vortex grit chambers. The process is called sedimentation.

Flow Equalization

Clarifiers and mechanized secondary treatment are more efficient under uniform flow conditions. Equalization basins may be used for temporary storage of diurnal or wet-weather flow peaks. Basins provide a place to temporarily hold incoming sewage during plant maintenance and a means of diluting and distributing batch discharges of toxic or high-strength waste which might otherwise inhibit biological secondary treatment (including portable toilet waste, vehicle holding tanks, and septic tank pumpers). Flow equalization basins require variable discharge control, typically include provisions for bypass and cleaning, and may also include aerators. Cleaning may be easier if the basin is downstream of screening and grit removal.

Fat and Grease Removal

In some larger plants, fat and grease are removed by passing the sewage through a small tank where skimmers collect the fat floating on the surface. Air blowers in the base of the tank may also be used to help recover the fat as a froth. Many plants, however, use primary clarifiers with mechanical surface skimmers for fat and grease removal.

Primary Treatment

Primary treatment tanks in Oregon, USA.

In the primary sedimentation stage, sewage flows through large tanks, commonly called "pre-settling basins", "primary sedimentation tanks" or "primary clarifiers". The tanks are used to settle sludge while grease and oils rise to the surface and are skimmed off. Primary settling tanks are usually equipped with mechanically driven scrapers that continually drive the collected sludge towards a hopper in the base of the tank where it is pumped to sludge treatment facilities. Grease and oil from the floating material can sometimes be recovered for saponification (soap making).

Secondary Treatment

Secondary treatment is designed to substantially degrade the biological content of the sewage which are derived from human waste, food waste, soaps and detergent. The majority of municipal plants treat the settled sewage liquor using aerobic biological processes. To be effective, the biota require both oxygen and food to live. The bacteria and protozoa consume biodegradable soluble organic contaminants (e.g. sugars, fats, organic short-chain carbon molecules, etc.) and bind much of the less soluble fractions into floc. Secondary treatment systems are classified as fixed-film or suspended-growth systems.

- **Fixed-film** or **attached growth** systems include trickling filters, bio-towers, and rotating biological contactors, where the biomass grows on media and the sewage passes over its surface. The fixed-film principle has further developed into Moving Bed Biofilm Reactors (MBBR) and Integrated Fixed-Film Activated Sludge (IFAS) processes. An MBBR system typically requires a smaller footprint than suspended-growth systems.
- **Suspended-growth** systems include activated sludge, where the biomass is mixed with the sewage and can be operated in a smaller space than trickling filters that treat the same amount of water. However, fixed-film systems are more able to cope with drastic changes in the amount of biological material and can provide higher removal rates for organic material and suspended solids than suspended growth systems.

Secondary Sedimentation

Secondary clarifier at a rural treatment plant.

Some secondary treatment methods include a secondary clarifier to settle out and separate biological floc or filter material grown in the secondary treatment bioreactor.

List of Process Types

- Activated sludge
- Aerated lagoon
- Aerobic granulation
- Constructed wetland
- Membrane bioreactor
- Rotating biological contactor
- Sequencing batch reactor
- Trickling filter

To use less space, treat difficult waste, and intermittent flows, a number of designs of hybrid treatment plants have been produced. Such plants often combine at least two stages of the three main treatment stages into one combined stage. In the UK, where a large number of wastewater treatment plants serve small populations, package plants are a viable alternative to building a large structure for each process stage. In the US, package plants are typically used in rural areas, highway rest stops and trailer parks.

Tertiary Treatment

The purpose of tertiary treatment is to provide a final treatment stage to further improve the effluent quality before it is discharged to the receiving environment (sea, river, lake, wet lands, ground, etc.). More than one tertiary treatment process may be used at any treatment plant. If disinfection is practised, it is always the final process. It is also called "effluent polishing."

Filtration

Sand filtration removes much of the residual suspended matter. Filtration over activated carbon, also called carbon adsorption, removes residual toxins.

Lagoons or ponds

Lagoons or ponds provide settlement and further biological improvement through storage in large man-made ponds or lagoons. These lagoons are highly aerobic and colonization by native macrophytes, especially reeds, is often encouraged. Small filter-feeding invertebrates such as Daphnia and species of Rotifera greatly assist in treatment by removing fine particulates.

A sewage treatment plant and lagoon in Everett, Washington, United States.

Biological Nutrient Removal

Biological nutrient removal (BNR) is regarded by some as a type of secondary treatment process, and by others as a tertiary (or "advanced") treatment process.

Wastewater may contain high levels of the nutrients nitrogen and phosphorus. Excessive release to the environment can lead to a buildup of nutrients, called eutrophication, which can in turn encourage the overgrowth of weeds, algae, and cyanobacteria (blue-green algae). This may cause an algal bloom, a rapid growth in the population of algae. The algae numbers are unsustainable and eventually most of them die. The decomposition of the algae by bacteria uses up so much of the oxygen in the water that most or all of the animals die, which creates more organic matter for the bacteria to decompose. In addition to causing deoxygenation, some algal species produce toxins that contaminate drinking water supplies. Different treatment processes are required to remove nitrogen and phosphorus.

Nitrogen Removal

Nitrogen is removed through the biological oxidation of nitrogen from ammonia to nitrate (nitrification), followed by denitrification, the reduction of nitrate to nitrogen gas. Nitrogen gas is released to the atmosphere and thus removed from the water.

Nitrification itself is a two-step aerobic process, each step facilitated by a different type of bacteria. The oxidation of ammonia (NH_3) to nitrite (NO_2^-) is most often facilitated by Nitrosomonas spp. ("nitroso referring to the formation of a nitroso functional group). Nitrite oxidation to nitrate (NO_3^-), though traditionally believed to be facilitated by Nitrobacter spp. (nitro referring the formation of a nitro functional group), is now known to be facilitated in the environment almost exclusively by Nitrospira spp.

Denitrification requires anoxic conditions to encourage the appropriate biological communities to form. It is facilitated by a wide diversity of bacteria. Sand filters, lagooning and reed beds can all be used to reduce nitrogen, but the activated sludge process (if designed well) can do the job the most easily.[17–18] Since denitrification is the reduction of nitrate to dinitrogen (molecular nitrogen) gas, an electron donor is needed. This can be, depending on the waste water, organic matter (from feces), sulfide, or an added donor like methanol. The sludge in the anoxic tanks (denitrification tanks) must be mixed well (mixture of recirculated mixed liquor, return activated sludge [RAS], and raw influent) e.g. by using submersible mixers in order to achieve the desired denitrification.

Sometimes the conversion of toxic ammonia to nitrate alone is referred to as tertiary treatment.

Over time, different treatment configurations have evolved as denitrification has become more sophisticated. An initial scheme, the Ludzack-Ettinger Process, placed an anoxic treatment zone before the aeration tank and clarifier, using the return activated sludge (RAS) from the clarifier as a nitrate source. Influent wastewater (either raw or as effluent from primary clarification) serves as the electron source for the facultative bacteria to metabolize carbon, using the inorganic nitrate as a source of oxygen instead of dissolved molecular oxygen. This denitrification scheme was naturally limited to the amount of soluble nitrate present in the RAS. Nitrate reduction was limited because RAS rate is limited by the performance of the clarifier.

The Modified Ludzak-Ettinger Process (MLE) is an improvement on the original concept, for it recycles mixed liquor from the discharge end of the aeration tank to the head of the anoxic tank to provide a consistent source of soluble nitrate for the facultative bacteria. In this instance, raw wastewater continues to provide the electron source, and sub-surface mixing maintains the bacteria in contact with both electron source and soluble nitrate in the absence of dissolved oxygen.

Many sewage treatment plants use centrifugal pumps to transfer the nitrified mixed liquor from the aeration zone to the anoxic zone for denitrification. These pumps are often referred to as Internal Mixed Liquor Recycle (IMLR) pumps. IMLR may be 200% to 400% the flow rate of influent wastewater (Q.) This is in addition to Return Activated Sludge (RAS) from secondary clarifiers, which may be 100% of Q. (Therefore, the hydraulic capacity of the tanks in such a system should handle at least 400% of annual average design flow (AADF.) At times, the raw or primary effluent wastewater must be carbon-supplemented by the addition of methanol, acetate, or simple food waste (molasses, whey, plant starch) to improve the treatment efficiency. These carbon additions should be accounted for in the design of a treatment facility's organic loading.

Further modifications to the MLE were to come: Bardenpho and Biodenipho processes include additional anoxic and oxidative processes to further polish the conversion of nitrate ion to molecular nitrogen gas. Use of an anaerobic tank following the initial anoxic process allows for luxury uptake of phosphorus by bacteria, thereby biologically reducing orthophosphate ion in the treated wastewater. Even newer improvements, such as Anammox Process, interrupt the formation of nitrate at the nitrite stage of nitrification, shunting nitrite-rich mixed liquor activated sludge to treatment where nitrite is then converted to molecular nitrogen gas, saving energy, alkalinity, and secondary carbon sourcing. Anammox™ (ANaerobic AMMonia OXidation) works by artificially extending detention time and preserving denitrifiying bacteria through the use of substrate added to the mixed liquor and continuously recycled from it prior to secondary clarification. Many other proprietary schemes are being deployed, including DEMON™, Sharon-ANAMMOX™, ANITA-Mox™, and DeAmmon™. The bacteria Brocadia anammoxidans can remove ammonium from waste water through anaerobic oxidation of ammonium to hydrazine, a form of rocket fuel.

Phosphorus Removal

Every adult human excretes between 200 and 1000 grams of phosphorus annually. Studies of United States sewage in the late 1960s estimated mean per capita contributions of 500 grams in urine and feces, 1000 grams in synthetic detergents, and lesser variable amounts used as corrosion and scale control chemicals in water supplies. Source control via alternative detergent formulations has subsequently reduced the largest contribution, but the content of urine and feces will remain unchanged. Phosphorus removal is important as it is a limiting nutrient for algae growth in many fresh water systems. (For a description of the negative effects of algae, see Nutrient removal). It is also particularly important for water reuse systems where high phosphorus concentrations may lead to fouling of downstream equipment such as reverse osmosis.

Phosphorus can be removed biologically in a process called enhanced biological phosphorus removal. In this process, specific bacteria, called polyphosphate-accumulating organisms (PAOs), are selectively enriched and accumulate large quantities of phosphorus within their cells (up to 20 percent of their mass). When the biomass enriched in these bacteria is separated from the treated water, these biosolids have a high fertilizer value.

Phosphorus removal can also be achieved by chemical precipitation, usually with salts of iron (e.g. ferric chloride), aluminum (e.g. alum), or lime.[18] This may lead to excessive sludge production as hydroxides precipitates and the added chemicals can be expensive. Chemical phosphorus removal requires significantly smaller equipment footprint than biological removal, is easier to operate and is often more reliable than biological phosphorus removal. Another method for phosphorus removal is to use granular laterite.

Once removed, phosphorus, in the form of a phosphate-rich sewage sludge, may be dumped in a landfill or used as fertilizer. In the latter case, the treated sewage sludge is also sometimes referred to as biosolids.

Disinfection

The purpose of disinfection in the treatment of waste water is to substantially reduce the number of microorganisms in the water to be discharged back into the environment for the later use of drinking, bathing, irrigation, etc. The effectiveness of disinfection depends on the quality of the water being treated (e.g., cloudiness, pH, etc.), the type of disinfection being used, the disinfectant dosage (concentration and time), and other environmental variables. Cloudy water will be treated less successfully, since solid matter can shield organisms, especially from ultraviolet light or if contact times are low. Generally, short contact times, low doses and high flows all militate against effective disinfection.

Common methods of disinfection include ozone, chlorine, ultraviolet light, or sodium hypochlorite.[16] Chloramine, which is used for drinking water, is not used in the treatment of waste water because of its persistence. After multiple steps of disinfection, the treated water is ready to be released back into the water cycle by means of the nearest body of water or agriculture. Afterwards, the water can be transferred to reserves for everyday human uses.

Chlorination remains the most common form of waste water disinfection in North America due to its low cost and long-term history of effectiveness. One disadvantage is that chlorination of residual organic material can generate chlorinated-organic compounds that may be carcinogenic or harmful to the environment. Residual chlorine or chloramines may also be capable of chlorinating organic material in the natural aquatic environment. Further, because residual chlorine is toxic to aquatic species, the treated effluent must also be chemically dechlorinated, adding to the complexity and cost of treatment.

Ultraviolet (UV) light can be used instead of chlorine, iodine, or other chemicals. Because no chemicals are used, the treated water has no adverse effect on organisms that later consume it, as may be the case with other methods. UV radiation causes damage to the genetic structure of bacteria, viruses, and other pathogens, making them incapable of reproduction. The key disadvantages of UV disinfection are the need for frequent lamp maintenance and replacement and the need for a highly treated effluent to ensure that the target microorganisms are not shielded from the UV radiation (i.e., any solids present in the treated effluent may protect microorganisms from the UV light). In the United Kingdom, UV light is becoming the most common means of disinfection because of the concerns about the impacts of chlorine in chlorinating residual organics in the wastewater and in chlorinating organics in the receiving water. Some sewage treatment systems in Canada and the US also use UV light for their effluent water disinfection.

Ozone (O_3) is generated by passing oxygen (O_2) through a high voltage potential resulting in a third oxygen atom becoming attached and forming O_3. Ozone is very unstable and reactive and oxidizes most organic material it comes in contact with, thereby destroying many pathogenic microorganisms. Ozone is considered to be safer than chlorine because, unlike chlorine which has to be stored on site (highly poisonous in the event of an accidental release), ozone is generated on-site as needed. Ozonation also produces fewer disinfection by-products than chlorination. A disadvantage of ozone disinfection is the high cost of the ozone generation equipment and the requirements for special operators.

Fourth Treatment Stage

Micropollutants such as pharmaceuticals, ingredients of household chemicals, chemicals used in small businesses or industries, environmental persistent pharmaceutical pollutant (EPPP) or pesticides may not be eliminated in the conventional treatment process (primary, secondary and tertiary treatment) and therefore lead to water pollution. Although concentrations of those substances and their decompostion products are quite low, there is still a chance to harm aquatic organisms. For pharmaceuticals, the following substances have been identified as "toxicologically relevant": substances with endocrine disrupting effects, genotoxic substances and substances that enhance the development of bacterial resistances. They mainly belong to the group of environmental persistent pharmaceutical pollutants. Techniques for elimination of micropollutants via a fourth treatment stage during sewage treatment are being tested in Germany, Switzerland and the Netherlands. However, since those techniques are still costly, they are not yet applied on a regular basis. Such process steps mainly consist of activated carbon filters that adsorb the micropollutants. Ozone can also be applied as an oxidative method. Also the use of enzymes such as the enzyme laccase is under investigation. A new concept which could provide an energy-efficient treatment of micropollutants could be the use of laccase secreting fungi cultivated at a wastewater treatment plant to degrade micropollutants and at the same time to provide enzymes at a cathode of a microbial biofuel cells. Microbial biofuel cells are investigated for their property to treat organic matter in wastewater.

To reduce pharmaceuticals in water bodies, also “source control” measures are under investigation, such as innovations in drug development or more responsible handling of drugs.

Odor Control

Odors emitted by sewage treatment are typically an indication of an anaerobic or “septic” condition. Early stages of processing will tend to produce foul-smelling gases, with hydrogen sulfide being most common in generating complaints. Large process plants in urban areas will often treat the odors with carbon reactors, a contact media with bio-slimes, small doses of chlorine, or circulating fluids to biologically capture and metabolize the noxious gases. Other methods of odor control exist, including addition of iron salts, hydrogen peroxide, calcium nitrate, etc. to manage hydrogen sulfide levels.

High-density solids pumps are suitable for reducing odors by conveying sludge through hermetic closed pipework.

Energy Requirements

For conventional sewage treatment plants, around 30 percent of the annual operating costs is usually required for energy. The energy requirements vary with type of treatment process as well as wastewater load. For example, constructed wetlands have a lower energy requirement than activated sludge plants, as less energy is required for the aeration step. Sewage treatment plants that produce biogas in their sewage sludge treatment process with anaerobic digestion can produce enough energy to meet most of the energy needs of the sewage treatment plant itself.

In conventional secondary treatment processes, most of the electricity is used for aeration, pumping systems and equipment for the dewatering and drying of sewage sludge. Advanced wastewater treatment plants, e.g. for nutrient removal, require more energy than plants that only achieve primary or secondary treatment.

Sludge Treatment And Disposal

The sludges accumulated in a wastewater treatment process must be treated and disposed of in a safe and effective manner. The purpose of digestion is to reduce the amount of organic matter and the number of disease-causing microorganisms present in the solids. The most common treatment options include anaerobic digestion, aerobic digestion, and composting. Incineration is also used, albeit to a much lesser degree.

Sludge treatment depends on the amount of solids generated and other site-specific conditions. Composting is most often applied to small-scale plants with aerobic digestion for mid-sized operations, and anaerobic digestion for the larger-scale operations.

The sludgc is sometimes passed through a so-called pre-thickener which de-waters the sludge. Types of pre-thickeners include centrifugal sludge thickeners rotary drum sludge thickeners and belt filter presses. Dewatered sludge may be incinerated or transported offsite for disposal in a landfill or use as an agricultural soil amendment.

Environment Aspects

The outlet of the Karlsruhe sewage treatment plant flows into the Alb.

Many processes in a wastewater treatment plant are designed to mimic the natural treatment processes that occur in the environment, whether that environment is a natural water body or the ground. If not overloaded, bacteria in the environment will consume organic contaminants, although this will reduce the levels of oxygen in the water and may significantly change the overall ecology of the receiving water. Native bacterial populations feed on the organic contaminants, and the numbers of disease-causing microorganisms are reduced by natural environmental conditions such as predation or exposure to ultraviolet radiation. Consequently, in cases where the receiving environment provides a high level of dilution, a high degree of wastewater treatment may not be required. However, recent evidence has demonstrated that very low levels of specific contaminants in wastewater, including hormones (from animal husbandry and residue from human hormonal contraception methods) and synthetic materials such as phthalates that mimic hormones in their action, can have an unpredictable adverse impact on the natural biota and potentially on humans if the water is re-used for drinking water. In the US and EU, uncontrolled discharges of wastewater to the environment are not permitted under law, and strict water quality requirements are to be met, as clean drinking water is essential. (For requirements in the US, see Clean Water Act.) A significant threat in the coming decades will be the increasing uncontrolled discharges of wastewater within rapidly developing countries.

Effects on Biology

Sewage treatment plants can have multiple effects on nutrient levels in the water that the treated sewage flows into. These nutrients can have large effects on the biological life in the water in contact with the effluent. Stabilization ponds (or sewage treatment ponds) can include any of the following:

- Oxidation ponds, which are aerobic bodies of water usually 1–2 meters in depth that receive effluent from sedimentation tanks or other forms of primary treatment.
- Dominated by algae
- Polishing ponds are similar to oxidation ponds but receive effluent from an oxidation pond or from a plant with an extended mechanical treatment.
- Dominated by zooplankton
- Facultative lagoons, raw sewage lagoons, or sewage lagoons are ponds where sewage is added with no primary treatment other than coarse screening. These ponds provide effective treatment when the surface remains aerobic; although anaerobic conditions may develop near the layer of settled sludge on the bottom of the pond.[:552–554]
- Anaerobic lagoons are heavily loaded ponds.
- Dominated by bacteria
- Sludge lagoons are aerobic ponds, usually 2 to 5 meters in depth, that receive anaerobically digested primary sludge, or activated secondary sludge under water.
- Upper layers are dominated by algae

Phosphorus limitation is a possible result from sewage treatment and results in flagellate-dominated plankton, particularly in summer and fall.

A phytoplankton study found high nutrient concentrations linked to sewage effluents. High nutrient concentration leads to high chlorophyll a concentrations, which is a proxy for primary production in marine environments. High primary production means high phytoplankton populations and most likely high zooplankton populations, because zooplankton feed on phytoplankton. However, effluent released into marine systems also leads to greater population instability.

The planktonic trends of high populations close to input of treated sewage is contrasted by the bacterial trend. In a study of Aeromonas spp. in increasing distance from a wastewater source, greater change in seasonal cycles was found the furthest from the effluent. This trend is so strong that the furthest location studied actually had an inversion of the Aeromonas spp. cycle in comparison to that of fecal coliforms. Since there is a main pattern in the cycles that occurred simultaneously at all stations it indicates seasonal factors (temperature, solar radiation, phytoplankton) control of the bacterial population. The effluent dominant species changes from Aeromonas caviae in winter to Aeromonas sobria in the spring and fall while the inflow dominant species is Aeromonas caviae, which is constant throughout the seasons.

Treated Sewage Reuse

With suitable technology, it is possible to reuse sewage effluent for drinking water, although this is usually only done in places with limited water supplies, such as Windhoek and Singapore.

In Israel, about 50 percent of agricultural water use (total use was 1 billion cubic metres in 2008) is provided through reclaimed sewer water. Future plans call for increased use of treated sewer water as well as more desalination plants.

Sewage Treatment In Developing Countries

Few reliable figures exist on the share of the wastewater collected in sewers that is being treated in the world. A global estimate by UNDP and UN-Habitat is that 90% of all wastewater generated is released into the environment untreated. In many developing countries the bulk of domestic and industrial wastewater is discharged without any treatment or after primary treatment only.

In Latin America about 15 percent of collected wastewater passes through treatment plants (with varying levels of actual treatment). In Venezuela, a below average country in South America with respect to wastewater treatment, 97 percent of the country's sewage is discharged raw into the environment. In Iran, a relatively developed Middle Eastern country, the majority of Tehran's population has totally untreated sewage injected to the city's groundwater. However, the construction of major parts of the sewage system, collection and treatment, in Tehran is almost complete, and under development, due to be fully completed by the end of 2012. In Isfahan, Iran's third largest city, sewage treatment was started more than 100 years ago.

Only few cities in sub-Saharan Africa have sewer-based sanitation systems, let alone wastewater treatment plants, an exception being South Africa and – until the late 1990s- Zimbabwe. Instead, most urban residents in sub-Saharan Africa rely on on-site sanitation systems without sewers, such as septic tanks and pit latrines, and faecal sludge management in these cities is an enormous challenge.

Basic sewer systems were used for waste removal in ancient Mesopotamia, where vertical shafts carried the waste away into cesspools. Similar systems existed in the Indus Valley civilization in modern-day India and in Ancient Crete and Greece. In the Middle Ages the sewer systems built by the Romans fell into disuse and waste was collected into cesspools that were periodically emptied by workers known as 'rakers' who would often sell it as fertilizer to farmers outside the city.

The Great Stink of 1858 stimulated research into the problem of sewage treatment. In this caricature in The Times, Michael Faraday reports to Father Thames on the state of the river.

Modern sewage systems were first built in the mid-nineteenth century as a reaction to the exacerbation of sanitary conditions brought on by heavy industrialization and urbanization. Due to the contaminated water supply, cholera outbreaks occurred in 1832, 1849 and 1855 in London, killing tens of thousands of people. This, combined with the Great Stink of 1858, when the smell of untreated human waste in the River Thames became overpowering, and the report into sanitation reform of the Royal Commissioner Edwin Chadwick, led to the Metropolitan Commission of Sewers appointing Sir Joseph Bazalgette to construct a vast underground sewage system for the safe removal of waste. Contrary to Chadwick's recommendations, Bazalgette's system, and others later built in Continental Europe, did not pump the sewage onto farm land for use as fertilizer; it was simply piped to a natural waterway away from population centres, and pumped back into the environment.

Early Attempts

One of the first attempts at diverting sewage for use as a fertilizer in the farm was made by the cotton mill owner James Smith in the 1840s. He experimented with a piped distribution system initially proposed by James Vetch that collected sewage from his factory and pumped it into the outlying farms, and his success was enthusiastically followed by Edwin Chadwick and supported by organic chemist Justus von Liebig.

The idea was officially adopted by the Health of Towns Commission, and various schemes (known as sewage farms) were trialled by different municipalities over the next 50 years. At first, the heavier solids were channeled into ditches on the side of the farm and were covered over when full, but soon flat-bottomed tanks were employed as reservoirs for the sewage; the earliest patent was taken out by William Higgs in 1846 for "tanks or reservoirs in which the contents of sewers and drains from cities, towns and villages are to be collected and the solid animal or vegetable matters therein contained, solidified and dried..." Improvements to the design of the tanks included the introduction of the horizontal-flow tank in the 1850s and the radial-flow tank in 1905. These tanks had to be manually de-sludged periodically, until the introduction of automatic mechanical de-sludgers in the early 1900s.

The precursor to the modern septic tank was the cesspool in which the water was sealed off to prevent contamination and the solid waste was slowly liquified due to anaerobic action; it was invented by L.H Mouras in France in the 1860s. Donald Cameron, as City Surveyor for Exeter patented an improved version in 1895, which he called a 'septic tank'; septic having the meaning of 'bacterial'. These are still in worldwide use, especially in rural areas unconnected to large-scale sewage systems.

Chemical Treatment

Sir Edward Frankland, a distinguished chemist, who demonstrated the possibility of chemically treating sewage in the 1870s.

It was not until the late 19th century that it became possible to treat the sewage by chemically breaking it down through the use of microorganisms and removing the pollutants. Land treatment was also steadily becoming less feasible, as cities grew and the volume of sewage produced could no longer be absorbed by the farmland on the outskirts.

Sir Edward Frankland conducted experiments at the Sewage Farm in Croydon, England, during the 1870s and was able to demonstrate that filtration of sewage through porous gravel produced a nitrified effluent (the ammonia was converted into nitrate) and that the filter remained unclogged over long periods of time. This established the then revolutionary possibility of biological treatment of sewage using a contact bed to oxidize the waste. This concept was taken up by the chief chemist for the London Metropolitan Board of Works, William Libdin, in 1887:

...in all probability the true way of purifying sewage...will be first to separate the sludge, and then turn into neutral effluent... retain it for a sufficient period, during which time it should be fully aerated, and finally discharge it into the stream in a purified condition. This is indeed what is aimed at and imperfectly accomplished on a sewage farm.

From 1885 to 1891 filters working on this principle were constructed throughout the UK and the idea was also taken up in the US at the Lawrence Experiment Station in Massachusetts, where Frankland's work was confirmed. In 1890 the LES developed a ‹trickling filter' that gave a much more reliable performance.

Contact beds were developed in Salford, Lancashire and by scientists working for the London City Council in the early 1890s. According to Christopher Hamlin, this was part of a conceptual revolution that replaced the philosophy that saw sewage purification as the prevention of decomposition with one that tried to facilitate the biological process that destroy sewage naturally.

Contact beds were tanks containing the inert substance, such as stones or slate, that maximized the surface area available for the microbial growth to break down the sewage. The sewage was held in the tank until it was fully decomposed and it was then filtered out into the ground. This method quickly became widespread, especially in the UK, where it was used in Leicester, Sheffield, Manchester and Leeds. The bacterial bed was simultaneously developed by Joseph Corbett as Borough Engineer in Salford and experiments in 1905 showed that his method was superior in that greater volumes of sewage could be purified better for longer periods of time than could be achieved by the contact bed.

The Royal Commission on Sewage Disposal published its eighth report in 1912 that set what became the international standard for sewage discharge into rivers; the ‹20:30 standard', which allowed 20 mg Biochemical oxygen demand and 30 mg suspended solid per litre.

REFERENCES

1. Tchobanoglous, George; Burton, Franklin L.; Stensel, H. David; Metcalf & Eddy, Inc. (2003). Wastewater Engineering: Treatment and Reuse (4th ed.). McGraw-Hill. ISBN 0-07-112250-8.
2. Metcalf & Eddy, Inc. (1972). Wastewater Engineering. New York: McGraw-Hill. ISBN 0-07-041675-3.
3. Burrian, Steven J., et al. (1999).The Historical Development of Wet-Weather Flow Management. US Environmental Protection Agency (EPA). National Risk

Management Research Laboratory, Cincinnati, OH. Document No. EPA/600/ JA-99/275.

4. Burton, Jr., G. Allen; Pitt, Robert E. (2001). Chapter 2. Receiving Water Uses, Impairments, and Sources of Stormwater Pollutants. Stormwater Effects Handbook: A Toolbox for Watershed Managers, Scientists, and Engineers. New York: CRC/Lewis Publishers. ISBN 0-87371-924-7.
5. Khopkar, S. M. (2004). Environmental Pollution Monitoring And Control. New Delhi: New Age International. p. 299. ISBN 81-224-1507-5.
6. Water and Environmental Health at London and Loughborough (1999). "Waste water Treatment Options.Technical brief no. 64. London School of Hygiene & Tropical Medicine and Loughborough University.
7. EPA. Washington, DC (2004). Primer for Municipal Waste water Treatment Systems. Document no. EPA 832-R-04-001.
8. Chapter 3. Flow Equalization. Process Design Manual for Upgrading Existing Wastewater Treatment Plants (Report). EPA. October 1971.
9. Huber Company, Berching, Germany (2012). Sedimentation Tanks.
10. Barwal, Anjali; Chaudhary, Rubina (2014). To study the performance of biocarriers in moving bed biofilm reactor (MBBR) technology and kinetics of biofilm for retrofitting the existing aerobic treatment systems: a review. Reviews in Environmental Science and Bio/Technology. Springer. **13** *(3): 285–299.* doi:10.1007/s11157-014-9333-7.
11. Randall, Clifford W.; Sen, Dipankar (1996). Full-scale evaluation of an integrated fixed-film activated sludge (IFAS) process for enhanced nitrogen removal. Water Science and Technology. Elsevier. **33** *(12): 155–162.* doi:10.1016/0273-1223(96)00469-6.
12. IFAS/MBBR Sustainable Wastewater Treatment Solutions (PDF). Black & Veatch, Inc. 2009. Archived from the original (PDF) on 2010-12-14. Brochure.
13. EPA. Washington, DC (2000). "Package Plants.Wastewater Technology Fact Sheet. Document no. EPA 832-F-00-016.
14. http://www.wefnet.org/mopnew/Operation_of_Municipal_Wastewater_Treatment_Plants/Chapter%2022%20Revised_6th%20Edition.pdf
15. Chapter 3. Biological Treatment Processes. Emerging Technologies for Wastewater Treatment and In-Plant Wet Weather Management (Report). EPA. March 2013. EPA 832-R-12-011.
16. B. Kartal, G.J. Kuenen and M.C.M van Loosdrecht, Sewage Treatment with Anammox, Science, 2010, vol 328 p 702-3
17. http://news.nationalgeographic.com/news/2005/11/1109_051109_rocketfuel.html
18. Harhangi, H. R.; Le Roy, M; Van Alen, T; Hu, B. L.; Groen, J; Kartal, B; Tringe, S. G.; Quan, Z. X.; Jetten, M. S.; Op Den Camp, H. J. (2012). Hydrazine synthase, a unique phylomarker with which to study the presence and biodiversity of anammox bacteria. Appl. Environ. Microbiol. **78** *(3): 752–8. doi:10.1128/ AEM.07113-11. PMC 3264106. PMID 22138989.*

19. Process Design Manual for Phosphorus Removal(Report). EPA. 1976. pp. 2–1. EPA 625/1-76-001a.

20. Das, Tapas K. (August 2001). Ultraviolet disinfection application to a wastewater treatment plant. Clean Technologies and Environmental Policy. Springer Berlin/ Heidelberg. **3** *(2): 69–80.* doi:10.1007/S100980100108.

21. Florida Department of Environmental Protection. Tallahassee, FL. "Ultraviolet Disinfection for Domestic Waste water." 2010-03-17.

22. UBA (Umweltbundesamt) (2014): Maßnahmen zur Verminderung des Eintrages von Mikroschadstoffen in die Gewässer. Texte 85/2014

23. Walz, A., Götz, K. (2014): Arzneimittelwirkstoffe im Wasserkreislauf. ISOE-Materialien zur Sozialen Ökologie Nr. 36 (in German)

24. Hernández Leal, L. (2010). Removal of micropollutants from grey water – Combining biological and physical/chemical processes. PhD thesis, Wageningen University

25. Ternes, T. (2009): Challenges for the future: emerging micropollutants in urban water cycle. BFG, Koblenz, Germany (presentation for a seminar at UNESCO-IHE, Delft)

26. Margot, J.; et al. (2013). Bacterial versus fungal laccase: potential for micropollutant degradation. AMB Express. **3**: 63. doi:10.1186/2191-0855-3-63.

27. Crude mushroom solution to degrade micropollutants and increase the performance of biofuel cells: https://www.biooekonomie-bw.de/en/articles/news/crude-mushroom-solution-to-degrade-micropollutants-and-increase-the-performance-of-biofuel-cells/

28. Logan, B.; Regan, J. (2006). Microbial Fuel Cells—Challenges and Applications. Environmental Science & Technology. **40** *(17): 5172–5180.* doi:10.1021/es0627592.

29. Lienert, J.; Bürki, T.; Escher, B.I. (2007). Reducing micropollutants with source control: Substance flow analysis of 212 pharmaceuticals in faeces and urine. Water Science & Technology. **56** *(5): 87–96. doi:10.2166/wst.2007.560.*

30. Harshman, Vaughan; Barnette, Tony (2000-12-28). Wastewater Odor Control: An Evaluation of Technologies. Water Engineering & Management. ISSN 0273-2238.

31. Walker, James D. and Welles Products Corporation (1976)."Tower for removing odors from gases." U.S. Patent No. 4421534.

32. Hoffmann, H., Platzer, C., von Münch, E., Winker, M. (2011). Technology review of constructed wetlands – Subsurface flow constructed wetlands for greywater and domestic wastewater treatment. Deutsche Gesellschaft für Internationale Zusammenarbeit (GIZ) GmbH, Eschborn, Germany, p. 11

33. IPEC Consultants, Ltd., Burnaby, BC, Canada. "IFT – Internally-fed – rotary thickener." Accessed 2012-06-16.

34. Mine-engineer.com, Long Beach, CA. "Belt Filter Press, How Does It Work?" Accessed 2012-06-16.

35. GlobalSpec, Inc., East Greenbush, NY and Eco tubes.

36. "How to Select Dewatering Equipment." Accessed 2012-06-12.

37. Environment Agency (archive) – Persistent, bioaccumulative and toxic PBT substances at the Wayback Machine (archived August 4, 2006). environment-agency.gov.uk. Retrieved on 2012-12-19.

38. Natural Environmental Research Council – River sewage pollution found to be disrupting fish hormones. Planetearth.nerc.ac.uk. Retrieved on 2012-12-19.

39. Endocrine Disruption Found in Fish Exposed to Municipal Wastewater at the Wayback Machine (archived October 15, 2011). USGS

40. Haughey, A. (1968). The Planktonic Algae of Auckland Sewage Treatment Ponds. New Zealand Journal of Marine and Freshwater Research. **2** *(4): 721–766. doi:10.1080/00288330.1968.9515271.*

41. Edmondson, W.T. (1972). Nutrients and Phytoplankton in Lake Washington. in Nutrients and Eutrophication: The Limiting Nutrient Controversy. American Society of Limnology and Oceanography, Special Symposia. Vol. 1.

42. Caperon, J.; Cattell, S. A. & Krasnick, G. (1971). Phytoplankton Kinetics in a Subtropical Estuary: Eutrophication (PDF). Limnology and Oceanography. **16** *(4): 599–607. doi:10.4319/lo.1971.16.4.0599.*

43. Monfort, P; Baleux, B (1990). Dynamics of Aeromonas hydrophila, Aeromonas sobria, and Aeromonas caviae in a sewage treatment pond. Applied and Environmental Microbiology. **56** *(7): 1999–2006. PMC 184551. PMID 2389929.*

44. PUB (Singapore National Water Agency)(2011). NEWater: History.

45. Martin, Andrew (2008-08-10). Farming in Israel, without a drop to spare. New York Times.

46. Corcoran, E., C. Nellemann, E. Baker, R. Bos, D. Osborn, H. Savelli (eds) (2010). Sick water? : the central role of wastewater management in sustainable development : a rapid response assessment (PDF). Arendal, Norway: UNEP/GRID-Arendal. ISBN 978-82-7701-075-5.

47. Caribbean Environment Programme (1998). Appropriate Technology for Sewage Pollution Control in the Wider Caribbean Region (PDF). Kingston, Jamaica: United Nations Environment Programme. Retrieved 2009-10-12.Technical Report No. 40.

48. Massoud Tajrishy and Ahmad Abrishamchi (2005). Integrated Approach to Water and Wastewater Management for Tehran, Iran. Water Conservation, Reuse, and Recycling: Proceedings of the Iranian-American Workshop. Washington, D.C.: National Academies Press.

49. Zimbabwe: Water and Sanitation Crisis, Human Rights Watch, https://www.hrw.org/news/2013/11/19/zimbabwe-water-and-sanitation-crisis

50. Chowdhry, S., Koné, D. (2012). Business Analysis of Fecal Sludge Management: Emptying and Transportation Services in Africa and Asia – Draft final report. Bill & Melinda Gates Foundation, Seattle, USA

51. Ashton, John; Ubido, Janet (1991). The Healthy City and the Ecological Idea (PDF). Journal of the Society for the Social History of Medicine. **4** *(1): 173–181. doi:10.1093/*shm/4.1.173. Retrieved 8 July 2013.

52. Lewis Dunbar B. Gordon (1851). A short description of the plans of Captain James Vetch for the sewerage of the metropolis.

53. H. H. Stanbridge (1976). History of Sewage Treatment in Britain. Institute of Water Pollution Control.

54. P. F. Cooper. Historical aspects of wastewater treatment(PDF). Retrieved 2013-12-21.

55. Martin V. Melosi (2010). The Sanitary City: Environmental Services in Urban America from Colonial Times to the Present. University of Pittsburgh Press. p. 110. ISBN 9780822973379.

56. Colin A. Russell (2003). Edward Frankland: Chemistry, Controversy and Conspiracy in Victorian England. Cambridge University Press. pp. 372–380. ISBN 9780521545815.

57. Sharma, Sanjay Kumar; Sanghi, Rashmi (2012). Advances in Water Treatment and Pollution Prevention. Springer. ISBN 9789400742048. Retrieved 2013-02-07.

58. Epidemics, demonstration effects, and municipal investment in sanitation capital.

59. Edwin Chadwick and the Engineers, 1842–1854: Systems and Antisystems in the Pipe-and-Brick Sewers War Technology and Culture (PDF). 1992.

60. Tilley, David F. (2011). Aerobic Wastewater Treatment Processes: History and Development. IWA Publishing. ISBN 9781843395423. Retrieved 2013-02-07.

61. Final report of the commissioners appointed to inquire and report what methods of treating and disposing of sewage (1912). us.archive.org

CHAPTER 2

Biodegradable Polymers

Chapter Contents

Biodegradable polymers are a specific type of polymer that breaks down after its intended purpose to result in natural byproducts such as gases (CO_2, N_2), water, biomass, and inorganic salts. These polymers are found both naturally and synthetically made, and largely consist of ester, amide, and ether functional groups. Their properties and breakdown mechanism are determined by their exact structure. These polymers are often synthesized by condensation reactions, ring opening polymerization, and metal catalysts.

INTRODUCTION

Biodegradable polymers have a long history and since many are natural products, the precise timeline of their discovery and use cannot be accurately traced. One of the first medicinal uses of a biodegradable polymer was the catgut suture, which dates back to at least 100 AD. The first catgut sutures were made from the intestines of sheep, but modern catgut sutures are made from purified collagen extracted from the small intestines of cattle, sheep, or goats.

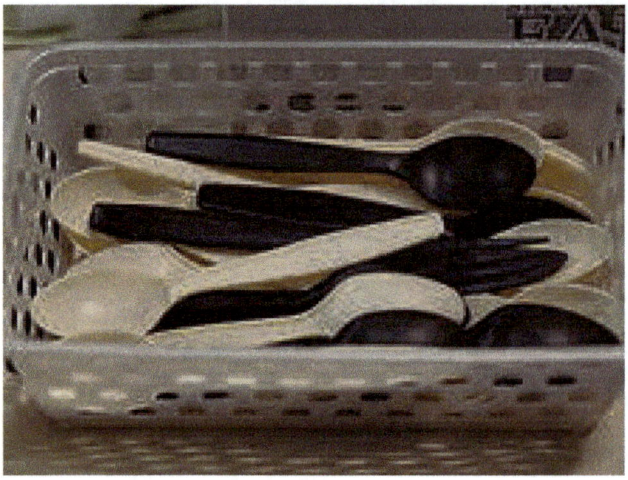

Example of cutlery made from biodegradable plastic.

The concept of synthetic biodegradable plastics and polymers was first introduced in the 1980s. In 1992, an international meeting was called where leaders in biodegradable polymers met to discuss a definition, standard, and testing protocol for biodegradable polymers. Also, oversight organizations such as American Society for Testing of Materials (ASTM) and the International Standards Organization (ISO) were created. Large clothing and grocery store chains have been making a push to utilize biodegradable bags in the late 2010s. Biodegradable polymers also received notice from various fields in 2012 when Professor Geoffrey Coates of Cornell University received the Presidential Green Chemistry Challenge Award. As of 2013, 5-10% of the plastic market focused on biodegradable polymer derived plastics.

STRUCTURE AND PROPERTIES

The structure of biodegradable polymers are instrumental in their properties. While there are innumerable biodegradable polymers, both synthetic and natural, there are a few commonalities among them.

Structure

Biodegradable polymers tend to consist of ester, amide, or ether bonds. In general, biodegradable polymers can be grouped into two large groups based on their structure and synthesis. One of these groups is agro-polymers, or those derived from biomass. The other consists of biopolyesters, which are those derived from microorganisms or synthetically made from either naturally or synthetic monomers.

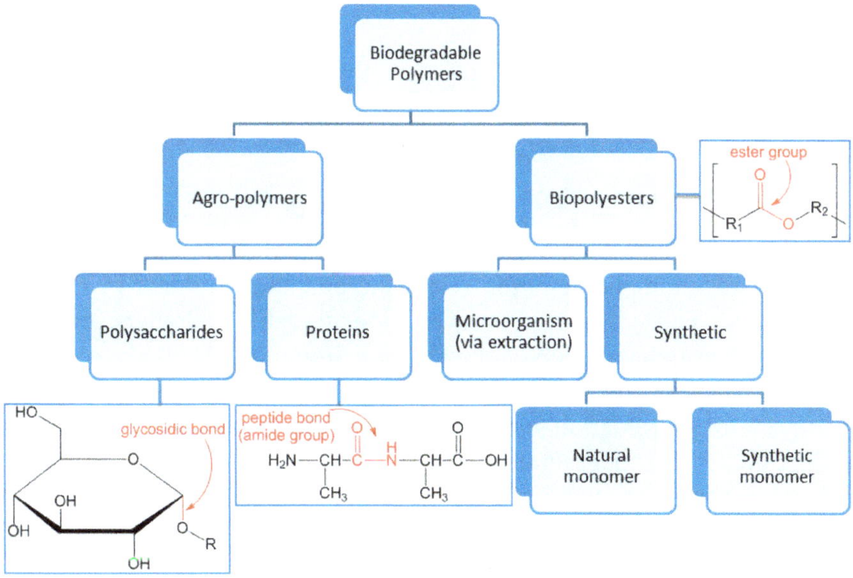

Biodegradable polymers organization based on structure and occurrence

Agro-polymers include polysaccharides, like starches found in potatoes or wood, and proteins, such as animal based whey or plant derived gluten. Polysacharides consist of glycosidic bonds, which take a hemiacetal of a saccharide and binds it to a alcohol via loss of water. Proteins are made from amino acids, which contain various functional groups. These amino acids come together again through condensation reactions to form peptide bonds, which consist of amide functional groups. Examples of biopolyesters includes polyhydroxybutyrate and polylactic acid.

Properties

Even though biodegradable polymers have numerous applications, there are properties that tend to be common among them. All biodegradable polymers should be stable and durable enough for use in their particular application, but upon disposal they should easily break down. Polymers, specifically biodegradable polymers, have extremely strong carbon backbones that are difficult to break, such that degradation often starts from the end-groups. Since the degradation begins at the end, a high surface area is common as it allows easy access for either the chemical, light, or organism. Biodegradable polymers also tend to have minimal chain branching as this cross linking often decreases the number of end groups per unit weight. Crystallinity is often low as it also inhibits access to end groups. A low degree of polymerization is normally seen, as hinted at above, as doing so allows for more accessible end groups for reaction with the degradation initiator. Another commonality of these polymers is their hydrophillicity. Hydrophobic polymers and end groups will prevent an enzyme from easily interacting if the water-soluble enzyme cannot easily get in contact with the polymer.

Other properties of biodegradable polymers that are common among those used for medicinal usages include

- non-toxic,
- capable of maintaining good mechanical integrity until degraded, and
- capable of controlled rates of degradation.

A goal is not to elicit the immune response, and the products of degradation also need not to be toxic. These are important as biodegradable polymers are used for drug delivery where it is critical to slowly release the drug into the body over time instead of all at once and that the pill is stable in the bottle until ready to be taken. Factors controlling the rate of degradation include: 1) percent crystallinity, 2) molecular weight, and 3) hydrophobicity. The degradation rate depends on the location in the body, which influences the environment surrounding the polymer such as pH, enzymes concentration, and amount of water among others. These are rapidly decomposed.

SYNTHESIS

One of the most important and most studied groups of biodegradable polymers are polyesters. Polyesters can be synthesized in a number of ways including direct condensation of alcohols and acids, ring opening polymerizations (ROP), and metal-catalyzed polymerization reactions. A great disadvantage of the step-wise polymerization via condensation of an acid and an alcohol is the need to continuously remove water from this system in order to drive the equilibrium of the reaction forward. This can necessitate harsh reaction conditions and long reaction times, resulting in a wide dispersity. A wide variety of starting materials can be used to synthesize polyesters, and each monomer type endows the final polymer chain with different characteristics and properties. The ROP of cyclic dimeric glycolic or lactic acid forms α-hydroxy acids which then polymerize into poly-(α-esters). A variety of organometallic initiators can be used to start the polymerization of polyesters, including tin, zinc, and aluminum complexes. The most common is tin(II) octanoate and has been approved as a food additive by the U.S. FDA, but there are still concerns about using the tin catalysts in the synthesis of biodegradable polymers for biomedical uses. The synthesis of poly(β-esters) and poly(γ-esters) can be carried out by similar ROP or condensation methods as with poly(γ-esters). Development of metal-free process that involve the use of bacterial or enzymatic catalysis in polyester formation is also being explored. These reactions have the benefit of generally being regioselective and stereospecific but suffer from the high cost of bacteria and enzymes, long reaction times, and products of low molecular weight.

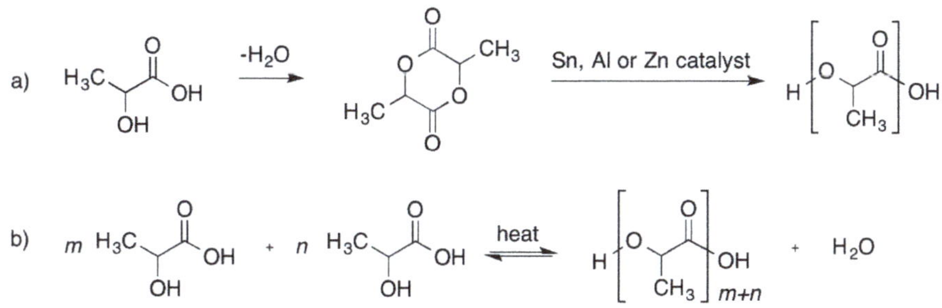

Example of routes to polyester formation using lactic acid. a) Condensation of lactic acid into dimeric lactide followed by ring-opening polymerization of to form poly(lactic acid); b) Direct condensation of lactic acid, demonstrating the need to continuously remove water from the system in order to drive the reaction forward.

While polyesters dominate both the research and industrial focus on synthetic biodegradable polymers, other classes of polymers are also of interest. Polyanhydrides are an active area of research in drug delivery because they only degrade from the surface and so are able to release the drug they carry at a constant rate. Polyanhydrides can be made via a variety of methods also used in the synthesis of other polymers, including condensation, dehydrochlorination, dehydrative coupling, and ROP. Polyurethanes and poly(ester amide)s are used in biomaterials. Polyurethanes were initially used for their biocompatibility, durability, resilience, but are more recently being investigated for their biodegradability. Polyurethanes are typically synthesized using a diisocyanate, a diol, and a polymer chain extender. The initial reaction is carried out between the diisocyanate and the diol, with the diisocyanate in excess to ensure that the ends of the new polymer chain are isocyanate groups. This polymer can then be reacted with either a diol or a diamine to form urethane or urethane-urea end groups, respectively. The choice of terminal groups affects the properties of the resulting polymer. Additionally, the use of vegetable oil and biomass in the formation of polyurethanes, as well as the conversion of polyurethanes to polyols, is an active area of research.

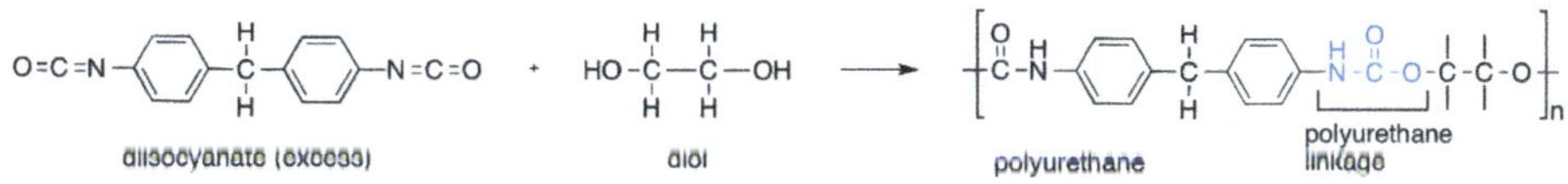

Synthesis of polyurethane from a diisocyanate and a diol. To cap this polymer, chain extenders of either diols or diamines can be added in order to tailor the properties.

MECHANISM OF BREAKDOWN

In general, biodegradable polymers break down to form gases, salts, and biomass. Complete biodegradation is said to occur when there are no oligomers or monomers left. The breakdown of these polymers depend on a variety of factors including the polymer and also, the environment the polymer is in. Polymer properties that influence degradation are bond type, solubility, and copolymers among others. The surrounding environment of the polymer is just as important as the polymer structure itself. These factors included items such as the pH, temperature, micoorganisms present, and water are just a few examples.

There are two primary mechanisms through which biodegradation can occur. One is through physical decomposition through reactions such as hydrolysis and photodegradation, which can lead to partial or complete degradation. The second mechanistic route is through biological processes which can be further broken down into aerobic and anaerobic processes. The first involves aerobic biodegradition, where oxygen is present and important. In this case, the general equation seen below where $C_{residue}$ represents smaller fragments of the initial polymer such as oligomers.

$$C_{polymer} + O_2 \longrightarrow C_{residue} + C_{biomass} + CO_2 + H_2O$$

General equation for aerobic biodegradition

The second mechanism of biodegradation is by anaerobic processes, where oxygen is not present.

$$C_{polymer} \longrightarrow C_{residue} + C_{biomass} + CO_2 + CH_4 + H_2O$$

General equation for anaerobic biodegradition

There are numerous organisms that have the ability to break down natural polymers. There are also synthetic polymers that have only been around for a hundred years with new features that microorganisms do not have the capability to break down. It will take millions of years before organisms can adapt to degrade all of these new synthetic polymers. Typically, after physical processes carry out the initial breakdown of the polymer, microorganisms will then take what is left and break down the components into even simpler units. These microorganisms normally take polymer fragments, such as oligomers or monomers, into the cell where enzymes work to make adenosine triphosphate (ATP) and polymer end products carbon dioxide, nitrogen gas, methane, water, minerals, and biomass.These enzymes act in a variety of ways to break down polymers including through oxidation or hydrolysis. Examples of key enzymes include proteases, esterases, glycosidases, and manganese peroxidases.

APPLICATIONS AND USES

Biodegradable polymers are of significant interest to a variety of fields including medicine, agriculture, and packaging. One of the most active areas of research in biodegradable polymer is in controlled drug delivery and release.

Medical

Biodegradable polymers have an innumerable uses in the biomedical field, particularly in the fields of tissue engineering and drug delivery. In order for a biodegradable polymer to be used as a therapeutic, it must meet several criteria: 1) be non-toxic in order to eliminate foreign body response; 2) the time it takes for the polymer to degrade is proportional to the time required for therapy; 3) the products resulting from biodegredation are not cytotoxic and are readily eliminated from the body; 4) the material must be easily processed in order to tailor the mechanical properties for the required task; 5) be easily sterilized; and 6) have acceptable shelf life.

Biodegradable polymers are of great interest in the field of drug delivery and nanomedicine. The great benefit of a biodegradable drug delivery system is the ability of the drug carrier to target the release of its payload to a specific site in the body and then degrade into nontoxic materials that are then eliminated from the body via natural metabolic pathways. The polymer slowly degrades into smaller fragments, releasing a natural product, and there is controlled ability to release a drug. The drug slowly releases as polymer degrades. For example, polylactic acid, poly(lactic-co-glycolic) acid, and poly(caprolactone), all of which are biodegradable, have been used to carry anti-cancer drugs. Encapsulating the therapeutic in a polymer and adding targeting agents decreases the toxicity of the drug to healthy cells.

Sutures made from polyglycolic acid. These sutures are absorbable and will be degraded by the body over time.

Biodegradable polymers and biomaterials are also of significant interest for tissue engineering and regeneration. Tissue engineering is the ability to regenerate tissue with the help of artificial materials. The perfection of such systems can be used to grow tissues and cells *in vitro* or use a biodegradable scaffold to construct new structures and organs *in vitro*. For these uses, a biodegradable scaffold is obviously preferred as it reduces the risk of immunological reaction and rejection of the foreign object. While many of the more advanced systems are not ready for human therapeutics, there is significant positive research in animal studies. For example, it was possible to successfully grow rat smooth muscle tissue on a polycaprolactone/polylactide scaffold. Further research and development may allow for this technology to be used for tissue replacement, support, or enhancement in humans. One of the ultimate goals of tissue engineering is the creation of organs, such as the kidney, from basic constituents. A scaffolding is necessary to grow the entity into a functioning organ, after which the polymer scaffold would degrade and be safely eliminated from the body. There are reports of using polyglycolic acid and polylactic acid to engineer vascular tissue for heart repair. The scaffold can be used to help create undamaged arteries and vessels.

In addition to tissue engineering, biodegradable polymers are being used orthopedic applications, such as bone and joint replacement. A wide variety of non-biodegradable polymers have been used for orthopedic applications including silicone rubber, polyethylene, acrylic resins, polyurethane, polypropylene, and polymethylmethacrylate. The primary role of many of these polymers was to act as a biocompatible cement in the fixation of prostheses and in the replacement of joints. Newer biologically compatible synthetic and natural biodegradable polymers have been developed; these include polyglycolide, polylactide, polyhydroxobutyrate, chitosan, hyaluronic acid, and hydrogels. In particular, poly(2-hydroxyethyl-methacrylate), poly(ethylene glycol), chitosan, and hyaluronic acid have been used extensively in the repair of cartilage, ligaments, and tendons. For example, poly(L-lactide) (PLA), is used to make screws and darts for meniscal repair and is marketed under the trade name Clearfix Mensical Dart/Screw. PLA is a slow degrading polymer and requires times greater than two years to degrade and be absorbed by the body.

Packaging and Materials

A trash bag made of a poly(lactic acid) blend, marketed under the brand Bio-Flex®

In addition to medicine, biodegradable polymers are often used to reduce the volume of waste in packaging materials.There is also significant effort to replace materials derived from petrochemicals with those that can be made from biodegradable components. One of the most commonly used polymers for packaging purposes is polylactic acid, PLA.The production of PLA has several advantages, the most important of which is the ability to tailor the physical properties of the polymer through processing methods. PLA is used for a variety of films, wrappings, and containers (including bottles and cups). In 2002, FDA ruled that PLA was safe to use in all food packaging. BASF markets a product called Ecovio® which is a blend of the company's biodegradable plastic Ecoflex® and PLA. An application for this biodegradable material is for thin plastic films such as shopping bags or trash bags.

NOTABLE EXAMPLES

2012 Presidential Green Chemistry Challenge

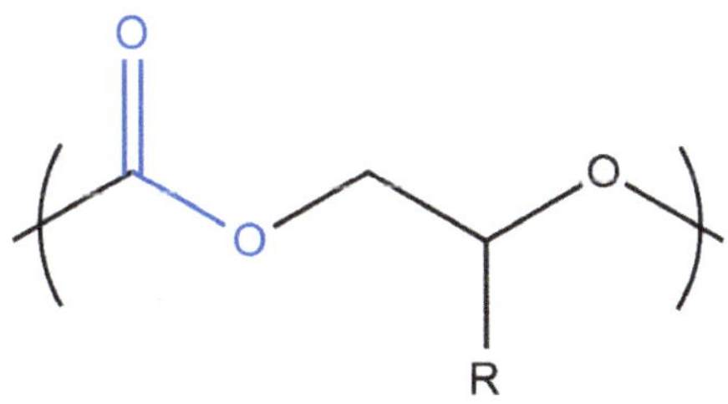

Carbon dioxide directly used in a polymer backbone.

Each year hundreds of millions of tons of plastics are produced from petroleum. Most of these plastics will remain in landfills for years to come or litter the environment posing significant health risks to animals; however, the average person's lifestyle would be impractical without them (see Applications). One solution to this conundrum lies in biodegradable polymers. These polymers have the distinct advantage that over time they will break down. Dr. Geoffrey Coates headed research to create catalysts that can not only efficiently create these biodegradable polymers, but the polymers also incorporate the greenhouse gas and global warming contributor, CO_2, and, environmentally present ground-ozone producer, CO. These two gases can be found or produced in high concentrations from agricultural waste, coal, and industrial applications as byproducts. Not only do the catalysts utilize these normally wasted and environmentally unfriendly gases, but they also do it extremely efficiently with high turnover numbers and frequencies in addition to good selectivity. These catalysts have been actively used by Novomer Inc to make polycarbonates that can replace the current coating bisphenol A (BPA) found in many food and drink packaging. Novomer's analysis shows that if used in all cases, these biodegradable polymer coatings could not only sequester, but also avoid further production of CO_2 in hundreds of millions of metric tons in just a single year.

REFERENCES

1. editors, Luc Avérous, Eric Pollet, (2012). Environmental silicate nano-biocomposites. London: Springer. ISBN 978-1-4471-4108-2.
2. Bastioli, editor, Catia (2005). Handbook of biodegradable polymers. Shawbury, Shrewsbury, Shropshire, U.K.: Rapra Technology. ISBN 9781847350442.
3. Nutton, Vivian (2012). Ancient medicine (2nd ed.). London: Routledge. ISBN 9780415520942.
4. editor, David B. Troy, (2005). Remington : The science and practice of pharmacy (21st ed.). Philadelphia, PA: Lippincott, Williams & Wilkins. ISBN 0-7817-4673-6.

5. Vroman, Isabelle; Tighzert, Lan (1 April 2009). Biodegradable Polymers . Materials. **2** *(2): 307–344.* doi:10.3390/ma2020307.

6. Kumar, A.Ashwin; K., Karthick (2011). Properties of Biodegradable Polymers and Degradation for Sustainable Development . International Journal of Chemical Engineering and Applications: 164–167. doi:10.7763/IJCEA.2011.V2.95.

7. Chamy, Rolando (June 14, 2013). Biodegradation - Life of Science. InTech. ISBN 978-953-51-1154-2.

8. Cox, David L. Nelson, Michael M. (2008). Lehninger principles of biochemistry (5th ed.). New York: W.H. Freeman. ISBN 978-0-7167-7108-1.

9. al.], edited by Buddy D. Ratner ... [et (2004). Biomaterials science : an introduction to materials in medicine (2nd ed.). San Diego, Calif.: Elsevier Academic Press. ISBN 0125824637.

10. Lendlein, edited by Andreas; Sisson, Adam (2011). Handbook of biodegradable polymers : synthesis, characterization and applications ([Elektronische Ressource] ed.). Weinheim: Wiley-VCH. ISBN 3527635831.

11. Amass, Wendy; Amass, Allan; Tighe, Brian (October 1998). A review of biodegradable polymers: uses, current developments in the synthesis and characterization of biodegradable polyesters, blends of biodegradable polymers and recent advances in biodegradation studies . Polymer International. **47** *(2): 89–144.* doi:10.1002/(SICI)1097-0126(1998100)47:2<89::AID-PI86>3.0.CO;2-F.

12. Brand, edited by Michael L. Johnson, Ludwig (2011). Computer methods. (1st ed.). San Diego, CA: Academic Press. ISBN 9781118164792.

13. Bastioli, ed.: Catia (2005). Handbook of biodegradable polymers (1. publ. ed.). Shawbury: Rapra Technology Ltd. ISBN 1-85957-389-4.

14. Martin, O; Avérous, L (June 2001). Poly(lactic acid): plasticization and properties of biodegradable multiphase systems . Polymer. **42** *(14): 6209–6219.* doi:10.1016/S0032-3861(01)00086-6.

15. Hollinger, edited by Jeffrey O. (2012). An introduction to biomaterials (2nd ed.). Boca Raton, FL: CRC Press/Taylor & Francis. ISBN 9781439812563.

16. Lligadas, Gerard; Ronda, Juan C.; Galià, Marina; Cádiz, Virginia (8 November 2010). Plant Oils as Platform Chemicals for Polyurethane Synthesis: Current State-of-the-Art . Biomacromolecules. **11** *(11): 2825–2835.* doi:10.1021/bm100839x.

17. Kržan, Andrej. Biodegradable polymers and plastics (PDF). Plastice. Retrieved 9 February 2014.

18. Tian, Huayu; Tang, Zhaohui; Zhuang, Xiuli; Chen, Xuesi; Jing, Xiabin (February 2012). Biodegradable synthetic polymers: Preparation, functionalization and biomedical application . Progress in Polymer Science. **37** *(2): 237–280. doi:10.1016/j.progpolymsci.2011.06.004.*

19. Middleton, John C; Tipton, Arthur J (December 2000). Synthetic biodegradable polymers as orthopedic devices . Biomaterials. **21** *(23): 2335–2346.* doi:10.1016/S0142-9612(00)00101-0.

20. Caballero-George, Catherina; Marin,; Briceño, (August 2013). Critical evaluation of biodegradable polymers used in nanodrugs . International Journal of Nanomedicine: 3071. doi:10.2147/IJN.S47186.

21. Bronzino, edited by Joon B. Park, Joseph D. (2002). Biomaterials Principles and Applications. Hoboken: CRC Press. ISBN 978-1-4200-4003-6.

22. Martina, Monique; Hutmacher, Dietmar W (February 2007). Biodegradable polymers applied in tissue engineering research: a review . Polymer International. **56** *(2): 145–157. doi:10.1002/pi.2108.*

23. Kurobe, H.; Maxfield, M. W.; Breuer, C. K.; Shinoka, T. (28 June 2012). Concise Review: Tissue-Engineered Vascular Grafts for Cardiac Surgery: Past, Present, and Future . Stem Cells Translational Medicine. **1** *(7): 566–571. doi:10.5966/sctm.2012-0044.*

24. Navarro, M; Michiardi, A; Castano, O; Planell, J.A (6 October 2008). Biomaterials in orthopaedics . Journal of The Royal Society Interface. **5** *(27): 1137–1158. doi:10.1098/rsif.2008.0151.*

25. Bio-Flex . Retrieved 10 February 2014.

26. Jamshidian, Majid; Tehrany, Elmira Arab; Imran, Muhammad; Jacquot, Muriel; Desobry, Stéphane (26 August 2010). Poly-Lactic Acid: Production, Applications, Nanocomposites, and Release Studies . Comprehensive Reviews in Food Science and Food Safety. **9** *(5): 552–571. doi:10.1111/j.1541-4337.2010.00126.x.*

27. FDA Food Contact Notification . Retrieved 10 February 2014.

28. BASF Ecovio . Retrieved 10 February 2014.

29. Plastics- The Facts 2012 (PDF). Plastics Europe. Retrieved 9 February 2014.

30. Winners of Presidential Green Chemistry Challenge Awards . American Chemical Society. Retrieved 9 February 2014.

31. 2012 Academic Award . United States Environmental Protection Agency. Retrieved 9 February 2014.

CHAPTER 3

Biofuels: Production and Use

Chapter Contents

Biofuel is fuel that contains recently dead matter as opposed to the ancient material that makes up fossil fuels. It can be made from materials such as plants, animal fat, oil, and sugar. The fuel can also be generated from specifically grown algae. It is non-toxic, renewable, and biodegradable.

When creating biofuel, alcohol, such as ethanol or methanol, is mixed with biological matter in order to cause the chemical reaction that gives it energy-generating properties. The fuel is typically combined with diesel fuel. The percentage of biological material in this kind of mixture can be anywhere from five to 50 percent of the fluid, depending on the practices of the region generating the fuel.

INTRODUCTION

A **biofuel** is a fuel that is produced through contemporary biological processes, such as agriculture and anaerobic digestion, rather than a fuel produced by geological processes such as those involved in the formation of fossil fuels, such as coal and petroleum, from prehistoric biological matter. Biofuels can be derived directly from plants, or indirectly from agricultural, commercial, domestic, and/or industrial wastes. Renewable biofuels generally involve contemporary carbon fixation, such as those that occur in plants or microalgae through the process of photosynthesis. Other renewable biofuels are made through the use or conversion of biomass (referring to recently living organisms, most often referring to plants or plant-derived materials). This biomass can be converted to convenient energy-containing substances in three different ways: thermal conversion, chemical conversion, and biochemical conversion. This biomass conversion can result in fuel in solid, liquid, or gas form. This new biomass can also be used directly for biofuels.

A bus fueled by biodiesel

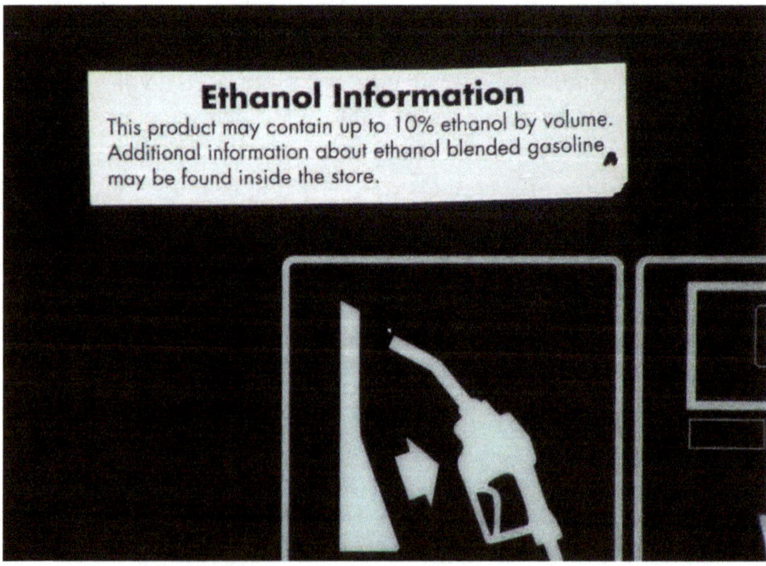

Information on pump regarding ethanol fuel blend up to 10%, California.

Bioethanol is an alcohol made by fermentation, mostly from carbohydrates produced in sugar or starch crops such as corn, sugarcane, or sweet sorghum. Cellulosic biomass, derived from non-food sources, such as trees and grasses, is also being developed as a feedstock for ethanol production. Ethanol can be used as a fuel for vehicles in its pure form, but it is usually used as a gasoline additive to increase octane and improve vehicle emissions. Bioethanol is widely used in the USA and in Brazil. Current plant design does not provide for converting the lignin portion of plant raw materials to fuel components by fermentation.

Biodiesel can be used as a fuel for vehicles in its pure form, but it is usually used as a diesel additive to reduce levels of particulates, carbon monoxide, and hydrocarbons from diesel-powered vehicles. Biodiesel is produced from oils or fats using transesterification and is the most common biofuel in Europe.

In 2010, worldwide biofuel production reached 105 billion liters (28 billion gallons US), up 17% from 2009, and biofuels provided 2.7% of the world's fuels for road transport. Global ethanol fuel production reached 86 billion liters (23 billion gallons US) in 2010, with the United States and Brazil as the world's top producers, accounting together for about 90% of global production. The world's largest biodiesel producer is the European Union, accounting for 53% of all biodiesel production in 2010. As of 2011, mandates for blending biofuels exist in 31 countries at the national level and in 29 states or provinces. The International Energy Agency has a goal for biofuels to meet more than a quarter of world demand for transportation fuels by 2050 to reduce dependence on petroleum and coal. The production of biofuels also led into a flourishing automotive industry, where by 2010, 79% of all cars produced in Brazil were made with a hybrid fuel system of bioethanol and gasoline.

There are various social, economic, environmental and technical issues relating to biofuels production and use, which have been debated in the popular media and scientific journals. These include: the effect of moderating oil prices, the food vs fuel debate, poverty reduction potential, carbon emissions levels, sustainable biofuel production, deforestation and soil erosion, loss of biodiversity, impact on water resources, rural social exclusion and injustice, shantytown migration, rural unskilled unemployment, and nitrous oxide (NO2) emissions.

LIQUID FUELS FOR TRANSPORTATION

Most transportation fuels are liquids, because vehicles usually require high energy density. This occurs naturally in liquids and solids. High energy density can also be provided by an internal combustion engine. These engines require clean-burning fuels. The fuels that are easiest to burn cleanly are typically liquids and gases. Thus, liquids meet the requirements of being both energy-dense and clean-burning. In addition, liquids (and gases) can be pumped, which means handling is easily mechanized, and thus less laborious.

First-Generation Biofuels

"First-generation" or conventional biofuels are made from sugar, starch, or vegetable oil.

Ethanol

Neat ethanol on the left (A), gasoline on the right (G) at a filling station in Brazil.

Biologically produced alcohols, most commonly ethanol, and less commonly propanol and butanol, are produced by the action of microorganisms and enzymes through the fermentation of sugars or starches (easiest), or cellulose (which is more difficult). Biobutanol (also called biogasoline) is often claimed to provide a direct replacement for gasoline, because it can be used directly in a gasoline engine.

Ethanol fuel is the most common biofuel worldwide, particularly in Brazil. Alcohol fuels are produced by fermentation of sugars derived from wheat, corn, sugar beets, sugar cane, molasses and any sugar or starch from which alcoholic beverages such as whiskey, can be made (such as potato and fruit waste, etc.). The ethanol production methods used are enzyme digestion (to release sugars from stored starches), fermentation of the sugars, distillation and drying. The distillation process requires significant energy input for heat (sometimes unsustainable natural gas fossil fuel, but cellulosic biomass such as bagasse, the waste left after sugar cane is pressed to extract its juice, is the most common fuel in Brazil, while pellets, wood chips and also waste heat are more common in Europe) Waste steam fuels ethanol factory - where waste heat from the factories also is used in the district heating grid.

Ethanol can be used in petrol engines as a replacement for gasoline; it can be mixed with gasoline to any percentage. Most existing car petrol engines can run on blends of up to 15% bioethanol with petroleum/gasoline. Ethanol has a smaller energy density than that of gasoline; this means it takes more fuel (volume and mass) to produce the same amount of work. An advantage of ethanol (CH_3CH_2OH) is that it has a higher octane rating than ethanol-free gasoline available at roadside gas stations, which allows an increase of an engine's compression ratio for increased thermal efficiency. In high-altitude (thin air) locations, some states mandate a mix of gasoline and ethanol as a winter oxidizer to reduce atmospheric pollution emissions.

U.S. President George W. Bush looks at sugar cane, a source of biofuel, with Brazilian President Luiz Inácio Lula da Silva during a tour on biofuel technology at Petrobras in São Paulo, Brazil, 9 March 2007.

Ethanol is also used to fuel bioethanol fireplaces. As they do not require a chimney and are flueless, bioethanol fires are extremely useful for newly built homes and apartments without a flue. The downsides to these fireplaces is that their heat output is slightly less than electric heat or gas fires, and precautions must be taken to avoid carbon monoxide poisoning.

Corn-to-ethanol and other food stocks has led to the development of cellulosic ethanol. According to a joint research agenda conducted through the US Department of Energy, the fossil energy ratios (FER) for cellulosic ethanol, corn ethanol, and gasoline are 10.3, 1.36, and 0.81, respectively.

Ethanol has roughly one-third lower energy content per unit of volume compared to gasoline. This is partly counteracted by the better efficiency when using ethanol (in a long-term test of more than 2.1 million km, the BEST project found FFV vehicles to be 1-26 % more energy efficient than petrol cars, but the volumetric consumption increases by approximately 30%, so more fuel stops are required.

With current subsidies, ethanol fuel is slightly cheaper per distance traveled in the United States.

Biodiesel

Biodiesel is the most common biofuel in Europe. It is produced from oils or fats using transesterification and is a liquid similar in composition to fossil/mineral diesel. Chemically, it consists mostly of fatty acid methyl (or ethyl) esters (FAMEs). Feedstocks for biodiesel include animal fats, vegetable oils, soy, rapeseed, jatropha, mahua, mustard, flax, sunflower, palm oil, hemp, field pennycress, *Pongamia pinnata* and algae. Pure biodiesel (B100) currently reduces emissions with up to 60% compared to diesel Second generation B100.

Biodiesel can be used in any diesel engine when mixed with mineral diesel. In some countries, manufacturers cover their diesel engines under warranty for B100 use, although Volkswagen of Germany, for example, asks drivers to check by telephone with the VW environmental services department before switching to B100. B100 may become more viscous at lower temperatures, depending on the feedstock used. In most cases, biodiesel is compatible with diesel engines from 1994 onwards, which use 'Viton' (by DuPont) synthetic rubber in their mechanical fuel injection systems. Note however, that no vehicles are certified for using neat biodiesel before 2014, as there was no emission control protocol available for biodiesel before this date.

Electronically controlled ‹common rail' and ‹unit injector' type systems from the late 1990s onwards may only use biodiesel blended with conventional diesel fuel. These engines have finely metered and atomized multiple-stage injection systems that are very sensitive to the viscosity of the fuel. Many current-generation diesel engines are made so that they can run on B100 without altering the engine itself, although this depends on the fuel rail design. Since biodiesel is an effective solvent and cleans residues deposited by mineral diesel, engine filters may need to be replaced more often, as the biofuel dissolves old deposits in the fuel tank and pipes. It also effectively cleans the engine combustion chamber of carbon deposits, helping to maintain efficiency. In many European countries, a 5% biodiesel blend is widely used and is available at thousands of gas stations. Biodiesel is also an oxygenated fuel, meaning it contains a reduced amount of carbon and higher hydrogen and oxygen content than fossil diesel. This improves the combustion of biodiesel and reduces the particulate emissions from unburnt carbon. However, using neat biodiesel may increase NOx-emissions

Biodiesel is also safe to handle and transport because it is non-toxic and biodegradable, and has a high flash point of about 300 °F (148 °C) compared to petroleum diesel fuel, which has a flash point of 125 °F (52 °C).

In the USA, more than 80% of commercial trucks and city buses run on diesel. The emerging US biodiesel market is estimated to have grown 200% from 2004 to 2005. By the end of 2006 biodiesel production was estimated to increase fourfold [from 2004] to more than 1 billion US gallons (3,800,000 m^3).

In France, biodiesel is incorporated at a rate of 8% in the fuel used by all French diesel vehicles. Avril Group produces under the brand Diester, a fifth of 11 million tons of biodiesel consumed annually by the European Union. It is the leading European producer of biodiesel.

Other Bioalcohols

Methanol is currently produced from natural gas, a non-renewable fossil fuel. In the future it is hoped to be produced from biomass as biomethanol. This is technically feasible, but the production is currently being postponed for concerns of Jacob S. Gibbs and Brinsley Coleberd that the economic viability is still pending. The methanol economy is an alternative to the hydrogen economy, compared to today's hydrogen production from natural gas.

Butanol (C_4H_9OH) is formed by ABE fermentation (acetone, butanol, ethanol) and experimental modifications of the process show potentially high net energy gains with butanol as the only liquid product. Butanol will produce more energy and allegedly can be burned "straight" in existing gasoline engines (without modification to the engine or car), and is less corrosive and less water-soluble than ethanol, and could be distributed via existing infrastructures. DuPont and BP are working together to help develop butanol. *E. coli* strains have also been successfully engineered to produce butanol by modifying their amino acid metabolism.

Green Diesel

Green diesel is produced through hydrocracking biological oil feedstocks, such as vegetable oils and animal fats. Hydrocracking is a refinery method that uses elevated temperatures and pressure in the presence of a catalyst to break down larger molecules, such as those found in vegetable oils, into shorter hydrocarbon chains used in diesel engines. It may also be called renewable diesel, hydrotreated vegetable oil or hydrogen-derived renewable diesel. Green diesel has the same chemical properties as petroleum-based diesel. It does not require new engines, pipelines or infrastructure to distribute and use, but has not been produced at a cost that is competitive with petroleum. Gasoline versions are also being developed. Green diesel is being developed in Louisiana and Singapore by ConocoPhillips, Neste Oil, Valero, Dynamic Fuels, and Honeywell UOP as well as Preem in Gothenburg, Sweden, creating what is known as Evolution Diesel.

Biofuel Gasoline

In 2013 UK researchers developed a genetically modified strain of Escherichia coli (E.Coli), which could transform glucose into biofuel gasoline that does not need to be blended. Later in 2013 UCLA researchers engineered a new metabolic pathway to bypass glycolysis and increase the rate of conversion of sugars into biofuel, while KAIST researchers developed a strain capable of producing short-chain alkanes, free fatty acids, fatty esters and fatty alcohols through the fatty acyl (acyl carrier protein (ACP)) to fatty acid to fatty acyl-CoA pathway *in vivo*. It is believed that in the future it will be possible to "tweak" the genes to make gasoline from straw or animal manure.

Vegetable Oil

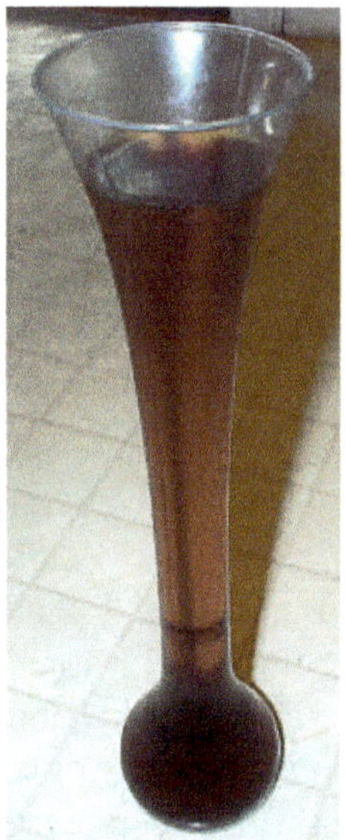

Filtered waste vegetable oil.

Walmart's truck fleet logs millions of miles each year, and the company planned to double the fleet's efficiency between 2005 and 2015. This truck is one of 15 based at Walmart's Buckeye, Arizona distribution center that was converted to run on a biofuel made from reclaimed cooking grease produced during food preparation at Walmart stores.

Straight unmodified edible vegetable oil is generally not used as fuel, but lower-quality oil can and has been used for this purpose. Used vegetable oil is increasingly being processed into biodiesel, or (more rarely) cleaned of water and particulates and used as a fuel.

As with 100% biodiesel (B100), to ensure the fuel injectors atomize the vegetable oil in the correct pattern for efficient combustion, vegetable oil fuel must be heated to reduce its viscosity to that of diesel, either by electric coils or heat exchangers. This is easier in warm or temperate climates. MAN B&W Diesel, Wärtsilä, and Deutz AG, as well as a number of smaller companies, such as Elsbett, offer engines that are compatible with straight vegetable oil, without the need for after-market modifications.

Vegetable oil can also be used in many older diesel engines that do not use common rail or unit injection electronic diesel injection systems. Due to the design of the combustion chambers in indirect injection engines, these are the best engines for use with vegetable oil. This system allows the relatively larger oil molecules more time to burn. Some older engines, especially Mercedes, are driven experimentally by enthusiasts without any conversion, a handful of drivers have experienced limited success with earlier pre-"Pumpe Duse VW TDI engines and other similar engines with direct injection. Several companies, such as Elsbett or Wolf, have developed professional conversion kits and successfully installed hundreds of them over the last decades.

Oils and fats can be hydrogenated to give a diesel substitute. The resulting product is a straight-chain hydrocarbon with a high cetane number, low in aromatics and sulfur and does not contain oxygen. Hydrogenated oils can be blended with diesel in all proportions. They have several advantages over biodiesel, including good performance at low temperatures, no storage stability problems and no susceptibility to microbial attack.

Bioethers

Bioethers (also referred to as fuel ethers or oxygenated fuels) are cost-effective compounds that act as octane rating enhancers." Bioethers are produced by the reaction of reactive iso-olefins, such as iso-butylene, with bioethanol.Bioethers are created by wheat or sugar beet. They also enhance engine performance, whilst significantly reducing engine wear and toxic exhaust emissions. Though bioethers are likely to replace petroethers in the UK, it is highly unlikely they will become a fuel in and of itself due to the low energy density. Greatly reducing the amount of ground-level ozone emissions, they contribute to air quality.

When it comes to transportation fuel there are six ether additives: dimethyl ether (DME), diethyl ether (DEE), methyl teritiary-butyl ether (MTBE), ethyl *ter*-butyl ether (ETBE), *ter*-amyl methyl ether (TAME), and *ter*-amyl ethyl ether (TAEE)

The European Fuel Oxygenates Association (EFOA) credits methyl Ttertiary-butyl ether (MTBE) and ethyl ter-butyl ether (ETBE) as the most commonly used ethers in fuel to replace lead. Ethers were introduced in Europe in the 1970s to replace the highly toxic compound. Although Europeans still use bio-ether additives, the US no longer has an oxygenate requirement therefore bio-ethers are no longer used as the main fuel additive.

Biogas

Pipes carrying biogas.

Biogas is methane produced by the process of anaerobic digestion of organic material by anaerobes. It can be produced either from biodegradable waste materials or by the use of energy crops fed into anaerobic digesters to supplement gas yields. The solid byproduct, digestate, can be used as a biofuel or a fertilizer.

Biogas can be recovered from mechanical biological treatment waste processing systems. Landfill gas, a less clean form of biogas, is produced in landfills through naturally occurring anaerobic digestion. If it escapes into the atmosphere, it is a potential greenhouse gas.

Farmers can produce biogas from manure from their cattle by using anaerobic digesters.

Syngas

Syngas, a mixture of carbon monoxide, hydrogen and other hydrocarbons, is produced by partial combustion of biomass, that is, combustion with an amount of oxygen that is not sufficient to convert the biomass completely to carbon dioxide and water. Before partial combustion, the biomass is dried, and sometimes pyrolysed. The resulting gas mixture, syngas, is more efficient than direct combustion of the original biofuel; more of the energy contained in the fuel is extracted.

Syngas may be burned directly in internal combustion engines, turbines or high-temperature fuel cells. The wood gas generator, a wood-fueled gasification reactor, can be connected to an internal combustion engine.

Syngas can be used to produce methanol, DME and hydrogen, or converted via the Fischer-Tropsch process to produce a diesel substitute, or a mixture of alcohols that can be blended into gasoline. Gasification normally relies on temperatures greater than 700 °C.

Lower-temperature gasification is desirable when co-producing biochar, but results in syngas polluted with tar.

Solid Biofuels

Examples include wood, sawdust, grass trimmings, domestic refuse, charcoal, agricultural waste, nonfood energy crops, and dried manure.

When raw biomass is already in a suitable form (such as firewood), it can burn directly in a stove or furnace to provide heat or raise steam. When raw biomass is in an inconvenient form (such as sawdust, wood chips, grass, urban waste wood, agricultural residues), the typical process is to densify the biomass. This process includes grinding the raw biomass to an appropriate particulate size (known as hogfuel), which, depending on the densification type, can be from 1 to 3 cm (0.4 to 1.2 in), which is then concentrated into a fuel product. The current processes produce wood pellets, cubes, or pucks. The pellet process is most common in Europe, and is typically a pure wood product. The other types of densification are larger in size compared to a pellet, and are compatible with a broad range of input feedstocks. The resulting densified fuel is easier to transport and feed into thermal generation systems, such as boilers.

Industry has used sawdust, bark and chips for fuel for decades, primary in the pulp and paper industry, and also bagasse (spent sugar cane) fueled boilers in the sugar cane industry. Boilers in the range of 500,000 lb/hr of steam, and larger, are in routine operation, using grate, spreader stoker, suspension burning and fluid bed combustion. Utilities generate power, typically in the range of 5 to 50 MW, using locally available fuel. Other industries have also installed wood waste fueled boilers and dryers in areas with low cost fuel.

One of the advantages of biomass fuel is that it is often a byproduct, residue or waste-product of other processes, such as farming, animal husbandry and forestry. In theory, this means fuel and food production do not compete for resources, although this is not always the case.

A problem with the combustion of raw biomass is that it emits considerable amounts of pollutants, such as particulates and polycyclic aromatic hydrocarbons. Even modern pellet boilers generate much more pollutants than oil or natural gas boilers. Pellets made from agricultural residues are usually worse than wood pellets, producing much larger emissions of dioxins and chlorophenols.

In spite of the above noted study, numerous studies have shown biomass fuels have significantly less impact on the environment than fossil based fuels. Of note is the US Department of Energy Laboratory, operated by Midwest Research Institute Biomass Power and Conventional Fossil Systems with and without CO2 Sequestration – Comparing the Energy Balance, Greenhouse Gas Emissions and Economics Study. Power generation emits significant amounts of greenhouse gases (GHGs), mainly carbon dioxide (CO_2). Sequestering CO_2 from the power plant flue gas can significantly reduce the GHGs from the power plant itself, but this is not the total picture. CO_2 capture and sequestration consumes additional energy, thus lowering the plant's fuel-to-electricity efficiency. To compensate for this, more fossil fuel must be procured and consumed to make up for lost capacity.

Taking this into consideration, the global warming potential (GWP), which is a combination of CO_2, methane (CH_4), and nitrous oxide (N_2O) emissions, and energy balance of the system need to be examined using a life cycle assessment. This takes into account the upstream processes which remain constant after CO_2sequestration, as well as the steps required for additional power generation. Firing biomass instead of coal led to a 148% reduction in GWP.

A derivative of solid biofuel is biochar, which is produced by biomass pyrolysis. Biochar made from agricultural waste can substitute for wood charcoal. As wood stock becomes scarce, this alternative is gaining ground. In eastern Democratic Republic of Congo, for example, biomass briquettes are being marketed as an alternative to charcoal to protect Virunga National Park from deforestation associated with charcoal production.

Second-Generation (advanced) Biofuels

Second generation biofuels, also known as advanced biofuels, are fuels that can be manufactured from various types of biomass. Biomass is a wide-ranging term meaning any source of organic carbon that is renewed rapidly as part of the carbon cycle. Biomass is derived from plant materials but can also include animal materials.

First generation biofuels are made from the sugars and vegetable oils found in arable crops, which can be easily extracted using conventional technology. In comparison, second generation biofuels are made from lignocellulosic biomass or woody crops, agricultural residues or waste, which makes it harder to extract the required fuel. A series of physical and chemical treatments might be required to convert lignocellulosic biomass to liquid fuels suitable for transportation.

Sustainable Biofuels

Biofuels in the form of liquid fuels derived from plant materials are entering the market, driven mainly by the perception that they reduce climate gas emissions, and also by factors such as oil price spikes and the need for increased energy security. However, many of the biofuels that are currently being supplied have been criticised for their adverse impacts on the natural environment, food security, and land use. In 2008, the Nobel-prize winning chemist Paul J. Crutzen published findings that the release of nitrous oxide (N_2O) emissions in the production of biofuels means that overall they contribute more to global warming than the fossil fuels they replace.

The challenge is to support biofuel development, including the development of new cellulosic technologies, with responsible policies and economic instruments to help ensure that biofuel commercialization is sustainable. Responsible commercialization of biofuels represents an opportunity to enhance sustainable economic prospects in Africa, Latin America and Asia.

According to the Rocky Mountain Institute, sound biofuel production practices would not hamper food and fibre production, nor cause water or environmental problems, and would enhance soil fertility. The selection of land on which to grow the feedstocks is a critical component of the ability of biofuels to deliver sustainable solutions. A key consideration is the minimisation of biofuel competition for prime cropland.

BIOFUELS BY REGION

Bio Diesel Powered Fast Attack Craft Of Indian Navy patrolling during IFR 2016.The green bands on the vessels are indicative of the fact that the vessels are powered by bio-diesel.

There are international organizations such as IEA Bioenergy, established in 1978 by the OECD International Energy Agency (IEA), with the aim of improving cooperation and information exchange between countries that have national programs in bioenergy research, development and deployment. The UN International Biofuels Forum is formed by Brazil, China, India, Pakistan, South Africa, the United States and the European Commission. The world leaders in biofuel development and use are Brazil, the United States, France, Sweden and Germany. Russia also has 22% of world's forest, and is a big biomass (solid biofuels) supplier. In 2010, Russian pulp and paper maker, Vyborgskaya Cellulose, said they would be producing pellets that can be used in heat and electricity generation from its plant in Vyborg by the end of the year. The plant will eventually produce about 900,000 tons of pellets per year, making it the largest in the world once operational.

Biofuels currently make up 3.1% of the total road transport fuel in the UK or 1,440 million litres. By 2020, 10% of the energy used in UK road and rail transport must come from renewable sources – this is the equivalent of replacing 4.3 million tonnes of fossil oil each year. Conventional biofuels are likely to produce between 3.7 and 6.6% of the energy needed in road and rail transport, while advanced biofuels could meet up to 4.3% of the UK's renewable transport fuel target by 2020.

AIR POLLUTION

Biomass § Environmental impact, and Ethanol_fuel § Air_pollution

Biofuels are different from fossil fuels in regard to greenhouse gases but are similar to fossil fuels in that biofuels contribute to air pollution. Burning produces airborne carbon particulates, carbon monoxide and nitrous oxides. The WHO estimates 3.7 million premature deaths worldwide in 2012 due to air pollution.Brazil burns significant amounts of ethanol biofuel. Gas chromatograph studies were performed of ambient air in São Paulo, Brazil, and compared to Osaka, Japan, which does not burn ethanol fuel. Atmospheric Formaldehyde was 160% higher in Brazil, and Acetaldehyde was 260% higher.

DEBATES REGARDING THE PRODUCTION AND USE OF BIOFUEL

There are various social, economic, environmental and technical issues with biofuel production and use, which have been discussed in the popular media and scientific journals. These include: the effect of moderating oil prices, the "food vs fuel" debate, food prices, poverty reduction potential, energy ratio, energy requirements, carbon emissions levels, sustainable biofuel production, deforestation and soil erosion, loss of biodiversity, impact on water resources, the possible modifications necessary to run the engine on biofuel, as well as energy balance and efficiency. The International Resource Panel, which provides independent scientific assessments and expert advice on a variety of resource-related themes, assessed the issues relating to biofuel use in its first report *Towards sustainable production and use of resources: Assessing Biofuels*. "Assessing Biofuels" outlined the wider and interrelated factors that need to be considered when deciding on the relative merits of pursuing one biofuel over another. It concluded that not all biofuels perform equally in terms of their impact on climate, energy security and ecosystems, and suggested that environmental and social impacts need to be assessed throughout the entire life-cycle.

Another issue with biofuel use and production is the US has changed mandates many times because the production has been taking longer than expected. The Renewable Fuel Standard (RFS) set by congress for 2010 was pushed back to at best 2012 to produce 100 million gallons of pure ethanol (not blended with a fossil fuel).

Current Research

Research is ongoing into finding more suitable biofuel crops and improving the oil yields of these crops. Using the current yields, vast amounts of land and fresh water would be needed to produce enough oil to completely replace fossil fuel usage. It would require twice the land area of the US to be devoted to soybean production, or two-thirds to be devoted to rapeseed production, to meet current US heating and transportation needs.

Specially bred mustard varieties can produce reasonably high oil yields and are very useful in crop rotation with cereals, and have the added benefit that the meal left over after the oil has been pressed out can act as an effective and biodegradable pesticide.

The NFESC, with Santa Barbara-based Biodiesel Industries, is working to develop biofuels technologies for the US navy and military, one of the largest diesel fuel users in the world. A group of Spanish developers working for a company called Ecofasa announced a new biofuel made from trash. The fuel is created from general urban waste which is treated by bacteria to produce fatty acids, which can be used to make biofuels.

Ethanol Biofuels (bioethanol)

As the primary source of biofuels in North America, many organizations are conducting research in the area of ethanol production. The National Corn-to-Ethanol Research Center (NCERC) is a research division of Southern Illinois University Edwardsville dedicated solely to ethanol-based biofuel research projects. On the federal level, the USDA conducts a large amount of research regarding ethanol production in the United States. Much of this research is targeted toward the effect of ethanol production on domestic food markets. A division of the U.S. Department of Energy, the National Renewable Energy Laboratory (NREL), has also conducted various ethanol research projects, mainly in the area of cellulosic ethanol.

Cellulosic ethanol commercialization is the process of building an industry out of methods of turning cellulose-containing organic matter into fuel. Companies, such as Iogen, POET, and Abengoa, are building refineries that can process biomass and turn it into bioethanol. Companies, such as Diversa, Novozymes, and Dyadic, are producing enzymes that could enable a cellulosic ethanol future. The shift from food crop feedstocks to waste residues and native grasses offers significant opportunities for a range of players, from farmers to biotechnology firms, and from project developers to investors.

As of 2013, the first commercial-scale plants to produce cellulosic biofuels have begun operating. Multiple pathways for the conversion of different biofuel feedstocks are being used. In the next few years, the cost data of these technologies operating at commercial scale, and their relative performance, will become available. Lessons learnt will lower the costs of the industrial processes involved.

In parts of Asia and Africa where drylands prevail, sweet sorghum is being investigated as a potential source of food, feed and fuel combined. The crop is particularly suitable for growing in arid conditions, as it only extracts one seventh of the water used by sugarcane. In India, and other places, sweet sorghum stalks are used to produce biofuel by squeezing the juice and then fermenting into ethanol.

A study by researchers at the International Crops Research Institute for the Semi-Arid Tropics (ICRISAT) found that growing sweet sorghum instead of grain sorghum could increase farmers incomes by US$40 per hectare per crop because it can provide fuel in addition to food and animal feed. With grain sorghum currently grown on over 11 million hectares (ha) in Asia and on 23.4 million ha in Africa, a switch to sweet sorghum could have a considerable economic impact.

Algae Biofuels

From 1978 to 1996, the US NREL experimented with using algae as a biofuels source in the "Aquatic Species Program. A self-published article by Michael Briggs, at the UNH Biofuels Group, offers estimates for the realistic replacement of all vehicular fuel with biofuels by using algae that have a natural oil content greater than 50%, which Briggs suggests can be grown on algae ponds at wastewater treatment plants. This oil-rich algae can then be extracted from the system and processed into biofuels, with the dried remainder further reprocessed to create ethanol. The production of algae to harvest oil for biofuels has not yet been undertaken on a commercial scale, but feasibility studies have been conducted to arrive at the above yield estimate. In addition to its projected high yield, algaculture — unlike crop-based biofuels — does not entail a decrease in food production, since it requires neither farmland nor fresh water. Many companies are pursuing algae bioreactors for various purposes, including scaling up biofuels production to commercial levels. Prof. Rodrigo E. Teixeira from the University of Alabama in Huntsville demonstrated the extraction of biofuels lipids from wet algae using a simple and economical reaction in ionic liquids.

Jatropha

Several groups in various sectors are conducting research on *Jatropha curcas*, a poisonous shrub-like tree that produces seeds considered by many to be a viable source of biofuels feedstock oil. Much of this research focuses on improving the overall per acre oil yield of Jatropha through advancements in genetics, soil science, and horticultural practices.

SG Biofuels, a San Diego-based jatropha developer, has used molecular breeding and biotechnology to produce elite hybrid seeds that show significant yield improvements over first-generation varieties. SG Biofuels also claims additional benefits have arisen from such strains, including improved flowering synchronicity, higher resistance to pests and diseases, and increased cold-weather tolerance.

Plant Research International, a department of the Wageningen University and Research Centre in the Netherlands, maintains an ongoing Jatropha Evaluation Project that examines the feasibility of large-scale jatropha cultivation through field and laboratory experiments. The Center for Sustainable Energy Farming (CfSEF) is a Los Angeles-based nonprofit research organization dedicated to jatropha research in the areas of plant science, agronomy, and horticulture. Successful exploration of these disciplines is projected to increase jatropha farm production yields by 200-300% in the next 10 years.

Fungi

A group at the Russian Academy of Sciences in Moscow, in a 2008 paper, stated they had isolated large amounts of lipids from single-celled fungi and turned it into biofuels in an economically efficient manner. More research on this fungal species, *Cunninghamella japonica*, and others, is likely to appear in the near future.The recent discovery of a variant of the fungus *Gliocladium roseum* (later renamed Ascocoryne sarcoides) points toward the production of so-called myco-diesel from cellulose. This organism was recently discovered in the rainforests of northern Patagonia, and has the unique capability of converting cellulose into medium-length hydrocarbons typically found in diesel fuel. Many other fungi that can degrade cellulose and other polymers have been observed to produce molecules that are currently being engineered using organisms from other kingdoms, suggesting that fungi may play a large role in the bio-production of fuels in the future (reviewed in).

Animal Gut Bacteria

Microbial gastrointestinal flora in a variety of animals have shown potential for the production of biofuels. Recent research has shown that TU-103, a strain of *Clostridium* bacteria found in Zebra feces, can convert nearly any form of cellulose into butanol fuel. Microbes in panda waste are being investigated for their use in creating biofuels from bamboo and other plant materials. There has also been substantial research into the technology of using the gut microbiomes of wood-feeding insects for the conversion of lignocellulotic material into biofuel.

Greenhouse Gas Emissions

Some scientists have expressed concerns about land-use change in response to greater demand for crops to use for biofuel and the subsequent carbon emissions. The payback period, that is, the time it will take biofuels to pay back the carbon debt they acquire due to land-use change, has been estimated to be between 100 and 1000 years, depending on the specific instance and location of land-use change. However, no-till practices combined with cover-crop practices can reduce the payback period to three years for grassland conversion and 14 years for forest conversion.

A study conducted in the Tocantis State, in northern Brazil, found that many families were cutting down forests in order to produce two conglomerates of oilseed plants, the J. curcas (JC group) and the R. communis (RC group). This region is composed of 15% Amazonian rainforest with high biodiversity, and 80% Cerrado forest with lower biodiversity. During the study, the farmers that planted the JC group released over 2193 Mg CO_2, while losing 53-105 Mg CO_2 sequestration from deforestation; and the RC group farmers released 562 Mg CO_2, while losing 48-90 Mg CO_2 to be sequestered from forest depletion. The production of these types of biofuels not only led into an increased emission of carbon dioxide, but also to lower efficiency of forests to absorb the gases that these farms were emitting. This has to do with the amount of fossil fuel the production of fuel crops involves. In addition, the intensive use of monocropping agriculture requires large amounts of water irrigation, as well as of fertilizers, herbicides and pesticides. This does not only lead to poisonous chemicals to disperse on water runoff, but also to the emission of nitrous oxide (NO_2) as a fertilizer byproduct, which is three hundred times more efficient in producing a greenhouse effect than carbon dioxide (CO_2).

Converting rainforests, peatlands, savannas, or grasslands to produce food crop–based biofuels in Brazil, Southeast Asia, and the United States creates a "biofuel carbon debt" by releasing 17 to 420 times more CO_2 than the annual greenhouse gas (GHG) reductions that these biofuels would provide by displacing fossil fuels. Biofuels made from waste biomass or from biomass grown on abandoned agricultural lands incur little to no carbon debt.

REFERENCES

1. What is biofuel? definition and meaning. BusinessDictionary.com. Retrieved 30 May 2015.
2. Biofuels Make a Comeback Despite Tough Economy. Worldwatch Institute. 2011-08-31. Retrieved 2011-08-31.
3. REN21 (2011). Renewables 2011: Global Status Report (PDF). pp. 13–14. Archived from the original (PDF) on 2011-09-05. Retrieved 2015-01-03.
4. Technology Roadmap, Biofuels for Transport (PDF). 2011.
5. Hall, Jeremy; Matos, Stelvia; Silvestre, Bruno; Martin, Michael (2011). Managing Technological and Social Uncertainties of Innovation: The Evolution of Brazilian Energy and Agriculture. Technological Forecasting and Social Change. 78: 1147–1157. doi:10.1016/j.techfore.2011.02.005.
6. Energikunskap | Lär dig mer om energi - E.ON
7. Bio ethanol fires information bio ethanol fireplace. (2009)
8. see "Breaking the Biological Barriers to Cellulosic Ethanol"
9. Brinkman, N. et al., "Well-to-Wheels Analysis of Advanced/Vehicle Systems", 2005.
10. Farrell, A.E.; et al. (2006). Ethanol can Contribute to Energy and Environmental Goals. Science. 311: 506–8. doi:10.1126/science.1121416. PMID 16439656.
11. Hammerschlag, R (2006). Ethanol's Energy Return on Investment: A Survey of the Literature 1999-Present. Environ. Sci. Technol. 40: 1744–50. doi:10.1021/es052024h.
12. Verdis Polaris Aura – second generation B100 – The advanced green one - Welcome to Perstorp - World leader in several sectors of the specialty chemicals market
13. ADM Biodiesel: Hamburg, Leer, Mainz. Biodiesel.de. Retrieved 2010-07-14.
14. RRI Limited for Biodiesel Filling Stations. Welcome to Biodiesel Filling Stations. Biodieselfillingstations.co.uk. Retrieved 2010-07-14.
15. http://www.vtt.fi/inf/pdf/technology/2012/T46.pdf Nylund.N-O & Koponen.K. 2013. Fuel and Technology Alternatives for Buses. Overall Energy Efficiency and Emission Performance. IEA Bioenergy Task 46]. Possibly the new emission standards Euro VI/EPA 10 will lead to reduced NOx-levels also when using B100.
16. Biofuels Facts. Hempcar.org. Retrieved 2010-07-14.

17. THE FUTURIST, Will Thurmond. July–August 2007
18. (Avril Group : Activity Report 2014, p. 58)
19. (EurObserv'ER 2014, p. 4)
20. Börjesson.P. et al. 2013, REPORT f3 2013:13, p 170
21. ButylFuel, LLC Main Page. Butanol.com. 2005-08-15. Retrieved 2010-07-14.
22. Evans, Jon (14 January 2008). Biofuels aim higher. Biofuels, Bioproducts and Biorefining (BioFPR). Retrieved 2008-12-03.
23. Brown, Robert; Jennifer Holmgren. Fast Pyrolysis and Bio-Oil Upgrading(PDF). Retrieved 15 March 2012.
24. Alternative & Advanced Fuels. US Department of Energy. Retrieved 7 March 2012.
25. Knothe, Gerhard (2010). Biodiesel and renewable diesel: A comparison. Progress in Energy and Combustion Science
26. Jessica, Ebert. Breakthroughs in Green Gasoline Production. Biomass Magazine. Retrieved 14 August 2012.
27. Albrecht, KO; Hallen, RT (March 2011). A Brief Literature Overview of Various Routes to Biorenewable Fuels from Lipids for the National Alliance of Advanced Biofuels and Bio-products NAAB Consortium (PDF). Prepared by the US Department of Energy
28. Summers, Rebecca (24 April 2013) Bacteria churn out first ever petrol-like biofuel New Scientist, Retrieved 27 April 2013
29. Bogorad, I. W.; Lin, T. S.; Liao, J. C. (2013). Synthetic non-oxidative glycolysis enables complete carbon conservation. Nature. 502: 693–697. doi:10.1038/nature12575.
30. Choi, Y. J.; Lee, S. Y. (2013). Microbial production of short-chain alkanes. Nature. 502: 571–4. doi:10.1038/nature12536. PMID 24077097.
31. Nishimoto, Alex (March 10, 2014). Walmart Debuts Turbine-Powered WAVE Semi Truck Prototype. Motor Trend.
32. Wal-Mart To Test Hybrid Trucks. Sustainable Business. February 3, 2009.
33. Evans, G. "Liquid Transport Biofuels - Technology Status Report", National Non-Food Crops Centre, 2008-04-14. Retrieved on 2009-05-11. Archived 19 September 2008 at the Wayback Machine.
34. Rock, Kerry; Maurice Korpelshoek (2007). Bioethers Impact on the Gasoline Pool. Digital Refining. Retrieved 15 February 2014.
35. Biofuels - Types of Biofuels - Bioethers
36. Biofuels - Types of Biofuels - Bioethers. Retrieved 30 May 2015.
37. Council Directive 85/536/EEC of 5 December 1985 on crude-oil savings through the use of substitute fuel components in petrol. Eur-lex.europa.eu. Retrieved 2010-07-14.
38. Sukla, Mirtunjay Kumar; Thallada Bhaskar; A.K. Jain; S.K. Singal; M.O. Garg.

[<> Bio-Ethers as Transportation Fuel: A Review] Check |url=value (help) (PDF). Indian Institute of Petroleum Dehradun. Retrieved 15 February 2014.

39. What are Bio-Ethers? (PDF). . The European Fuel Oxygenates Association.
40. Gasoline. Environmental Protection Agency.
41. Redman, G., The Andersons Centre. "Assessment of on-farm AD in the UK", National Non-Food Crops Centre, 2008-06-09. Retrieved on 2009-05-11. Archived 13 November 2010 at the Wayback Machine.
42. "BIOGAS: No bull, manure can power your farm." Farmers Guardian (25 September 2009): 12. General OneFile. Gale.
43. Electricity from wood through the combination of gasification and solid oxide fuel cells, Ph.D. Thesis by Florian Nagel, Swiss Federal Institute of Technology Zurich, 2008
44. Biomass and Alternate Energy Fuel Systems: An Engineering and Economic Guide
45. Frauke Urban and Tom Mitchell 2011. Climate change, disasters and electricity generation Archived 20 September 2012 at the Wayback Machine.. London: Overseas Development Institute and Institute of Development Studies
46. Briens, Cedric; Piskorz, Jan; Berruti, Franco (2008). Biomass Valorization for Fuel and Chemicals Production -- A Review. International Journal of Chemical Reactor Engineering. 6: R2. doi:10.2202/1542-6580.1674.
47. Threat to Great Apes Highlighted at Virunga Meeting. America.gov. Archived from the original on 28 August 2010. Retrieved 2010-07-14.
48. Ramirez, Jerome; Brown, Richard; Rainey, Thomas (1 July 2015). A Review of Hydrothermal Liquefaction Bio-Crude Properties and Prospects for Upgrading to Transportation Fuels. Energies. 8: 6765–6794. doi:10.3390/en8076765.
49. The potential and challenges of drop-in fuels (members only) | IEA Bioenergy Task 39 – Commercializing Liquid Biofuels. task39.sites.olt.ubc.ca. Retrieved 2015-09-10.
50. The Royal Society (January 2008). Sustainable biofuels: prospects and challenges, ISBN 978-0-85403-662-2, p. 61.
51. Gordon Quaiattini. Biofuels are part of the solution Canada.com, April 25, 2008. Retrieved December 23, 2009.
52. Crutzen, P. J.; Mosier, A. R.; Smith, K. A.; Winiwarter, W. (2008). N_2O release from agro-biofuel production negates global warming reduction by replacing fossil fuels". Atmos. Chem. Phys. 8: 389–395. doi:10.5194/acp-8-389-2008.
53. EPFL Energy Center (c2007). Roundtable on Sustainable BiofuelsRetrieved December 23, 2009.
54. Rocky Mountain Institute (2005). Winning the Oil Endgame p. 107. Retrieved December 23, 2009.
55. The Royal Society (2008). p. 2.
56. Growing Sustainable Biofuels: Common Sense on Biofuels, part 2World

Changing, March 12, 2008. Retrieved December 24, 2008.

57. IEA bioenergy. IEA bioenergy. Archived from the original on 26 May 2010. Retrieved 2010-07-14.

58. Press Conference Launching International Biofuels Forum. United Nations Department of Public Information. 2007-03-02. Retrieved 2008-01-15.

59. Greenpeace - The Russian Forests Archived 25 August 2010 at the Wayback Machine.

60. "World's Largest Pellet Plant to Start by Year-End". Moscow Times

61. UK falls short of biofuel targets for 2010/2011. Retrieved 30 May 2015.

62. National Non-Food Crops Centre. "Advanced Biofuels: The Potential for a UK Industry, NNFCC 11-011", Retrieved on 2011-11-17

63. http://www.who.int/indoorair/interventions/antiguamod21.pdf

64. WHO | Ambient (outdoor) air quality and health

65. Atmospheric alcohols and aldehydes concentrations measured in Osaka, Japan and in Sao Paulo, Brazil

66. Fletcher Jr., Robert J.; Bruce A Robertson; Jason Evans; Patrick J Doran; Janaki RR Alavalapati; Douglas W Schemske (2011). Biodiversity conservation in the era of biofuels: risks and opportunities. Frontiers in Ecology and the Environment. 9 (3): 161–168. doi:10.1890/090091. Retrieved 10 December 2013.

67. Cotton, Charles A. R.; Jeffrey S. Douglass; Sven De Causmaeker; Katharina Brinkert; Tanai Cardona; Andrea Fantuzzi; A. William Rutherford; James W. Murray (2015). Photosynthetic constraints on fuel from microbes. Frontiers in Bioengineering and Biotechnology. 3. doi:10.3389/fbioe.2015.00036. Retrieved 18 March 2015.

68. Publications - International Resource Panel. Archived from the original on 1 January 2016. Retrieved 30 May 2015.

69. Bracmort, Kelsi. Meeting the Renewable Fuel Standard (RFS) Mandate for Cellulosic Biofuels:Questions and Answers (PDF). Washington, DC: Congressional Research Service.

70. Mustard Hybrids for Low-Cost Biofuels and Organic Pesticides (PDF). Archived from the original (PDF) on 26 July 2011. Retrieved 2010-03-15.

71. Future Energies (2003-10-30). PORT HUENEME, Calif: U.S. Navy to Produce its Own Biofuels :: Future Energies :: The future of energy. Future Energies. Retrieved 2009-10-17.

72. Newsvine - Ecofasa turns waste to biofuels using bacteria. Lele.newsvine.com. 2008-10-18. Retrieved 2009-10-17.

73. Ethanol Research (2012-04-02). National Corn-to-Ethanol Research Center (NCERC). Ethanol Research. Archived from the original on 20 March 2012. Retrieved 2012-04-02.

74. American Coalition for Ethanol (2008-06-02). Responses to Questions from

Senator Bingaman (PDF). American Coalition for Ethanol. Archived from the original (PDF) on 4 October 2011. Retrieved 2012-04-02.

75. National Renewable Energy Laboratory (2 March 2007). Research Advantages: Cellulosic Ethanol (PDF). National Renewable Energy Laboratory. Archived from the original (PDF) on 25 January 2012. Retrieved 2012-04-02.
76. Pernick, Ron and Wilder, Clint (2007). The Clean Tech Revolution p. 96.
77. HLPE (2013). Biofuels and food security (PDF).
78. Sweet Sorghum : A New Smart Biofuel Crop. Agriculture Business Week. 30 June 2008. Archived from the original on 27 May 2015.
79. Sweet sorghum for food, feed and fuel New Agriculturalist, January 2008.
80. Sheehan, John; et al. (July 1998). A Look Back at the U. S. Department of Energy's Aquatic Species Program: Biofuels from Algae (PDF). National Renewable Energy Laboratory. Retrieved 16 June 2012.
81. Briggs, Michael (August 2004). Widescale Biodiesel Production from Algae. UNH Biodiesel Group (University of New Hampshire). Archived from the originalon 24 March 2006. Retrieved 2007-01-02.
82. Valcent Products Inc. Develops Clean Green Vertical Bio-Reactor. Valcent Products. Archived from the original on 18 June 2008. Retrieved 2008-07-09.
83. Technology: High Yield Carbon Recycling. GreenFuel Technologies Corporation. Archived from the original on 21 August 2007. Retrieved 2008-07-09.
84. R. E. Teixeira (2012). Energy-efficient extraction of fuel and chemical feedstocks from algae. Green Chemistry. 14 (2): 419–427. doi:10.1039/C2GC16225C.
85. B.N. Divakara; H.D. Upadhyaya; S.P. Wani; C.L. Laxmipathi Gowda (2010). Biology and genetic improvement of Jatropha curcas L.: A review. Applied Energy. 87 (3): 732–742. doi:10.1016/j.apenergy.2009.07.013.
86. Biofuels Digest (2011-05-16). Jatropha blooms again: SG Biofuels secures 250K acres for hybrids. Biofuels Digest. Retrieved 2012-03-08.
87. SG Biofuels (2012-03-08). Jmax Hybrid Seeds. SG Biofuels. Retrieved 2012-03-08.
88. Plant Research International (2012-03-08). JATROPT (Jatropha curcas): Applied and technical research into plant properties. Plant Research International. Retrieved 2012-03-08.
89. Biofuels Magazine (2011-04-11). Energy Farming Methods Mature, Improve. Biofuels Magazine. Retrieved 2012-03-08.
90. Sergeeva, Y. E.; Galanina, L. A.; Andrianova, D. A.; Feofilova, E. P. (2008). Lipids of filamentous fungi as a material for producing biodiesel fuel. Applied Biochemistry and Microbiology. 44 (5): 523–527. doi:10.1134/S0003683808050128.
91. Strobel, G.; Knighton, B.; Kluck, K.; Ren, Y.; Livinghouse, T.; Griffin, M.; Spakowicz, D.; Sears, J. (2008). The production of myco-diesel hydrocarbons and their derivatives by the endophytic fungus Gliocladium roseum (NRRL

50072). Microbiology (Reading, England). 154 (Pt 11): 3319–3328. doi:10.1099/mic.0.2008/022186-0. PMID 18957585.

92. Spakowicz, Daniel J.; Strobel, Scott A. (2015). Biosynthesis of hydrocarbons and volatile organic compounds by fungi: bioengineering potential. Applied microbiology and biotechnology. 99 (12): 4943–4951. doi:10.1007/s00253-015-6641-y. Retrieved 2016-02-22.

93. Kathryn Hobgood Ray (August 25, 2011). Cars Could Run on Recycled Newspaper, Tulane Scientists Say. Tulane University news webpage. Tulane University. Retrieved March 14, 2012.

94. Panda Poop Might Help Turn Plants Into Fuel. News.nationalgeographic.com. 2013-09-10. Retrieved 2013-10-02.

95. Sun, Jian-Zhong; Scharf, Michael E. (2010). Exploring and integrating cellulolytic systems of insects to advance biofuel technology. Insect Science. 17: 163–165. doi:10.1111/j.1744-7917.2010.01348.x.

96. Searchinger, Timothy; Ralph Heimlich; R.A. Houghton; Fengxia Dong; Amani Elobeid; Jacinto Fabiosa; Simla Tokgoz; Dermot Hayes; Tun-Hsiang Yu (2011). Use of U.S. Croplands for Biofuels Increases Greenhouse Gases Through Emissions from Land-Use Change. Science. pp. 1238–1240. doi:10.1126/science.1151861. Retrieved 8 November 2011.

97. Kim, Hyungtae; Seungdo Kim; Bruce E. Dale (2009). Biofuels, Land Use Change, and Greenhouse Gas Emissions: Some Unexplored Variables. Environmental Science. pp. 961–967. doi:10.1021/es802681k.

98. Alves Finco, Marcus V.; Doppler, Werner (2010). Bioenergy and Sustainable Development: The Dilemma of Food Security and Climate Change in the Brazilian Savannah. Energy for Sustainable Development. 12: 194–199. doi:10.1016/j.esd.2010.04.006.

99. Runge, Ford; Senauer, Benjamin (2007). How Biofuels Could Starve the Poor. Foreign Affairs. 86: 41–53. JSTOR 20032348.

100. fargione, Joseph; Jason Hill; David Tilman; Stephen Polasky; Peter Hawthorne (2008). Land Clearing and the Biofuel Carbon Debt. Science. pp. 1235–1238. doi:10.1126/science.1152747. Retrieved 12 November 2011.

101. The National Academies Press (2008). Water Issues of Biofuel Production Plants. The National Academies Press. Retrieved 18 June 2015.

CHAPTER 4

Compost Management

Chapter Contents

- **Introduction**
- **Terminology**
- **Ingredients**
- **Uses**
- **Composting Technologies**
- **Compost and Land-Filling**
- **Related Technologies**
- **History**
- **Composting Toilet**
- **Terminology**

Composting speeds the natural process whereby microorganisms break down complex organic matter into simple, nutritious topsoil. When this occurs on the forest floor, it's called mulching, but when people have a hand in the procedure, we call it composting. Composting requires water, air, and rotting material with a good nitrogen to carbon ratio.

Composting starts with the proper equipment. A container needs to be durable, with slots for air circulation, compartments for different ingredients, and a cover to keep out rain and snow. Some people like to mount their compost bin on an axle so they can rotate it and mix everything together. Containers should be kept on level ground away from sources of contamination, like pets or garbage.

INTRODUCTION

Compost is organic matter that has been decomposed and recycled as a fertilizer and soil amendment. Compost is a key ingredient in organic farming.

At the simplest level, the process of composting requires making a heap of wetted organic matter known as green waste (leaves, food waste) and waiting for the materials to break down into humus after a period of weeks or months. Modern, methodical composting is a multi-step, closely monitored process with measured inputs of water, air, and carbon- and nitrogen-rich materials. The decomposition process is aided by shredding the plant matter, adding water and ensuring proper aeration by regularly turning the mixture. Worms and fungi further break up the material. Bacteria requiring oxygen to function (aerobic bacteria) and fungi manage the chemical process by converting the inputs into heat, carbon dioxide and ammonium. The ammonium (NH_4) is the form of nitrogen used by plants. When available ammonium is not used by plants it is further converted by bacteria into nitrates (NO_3) through the process of nitrification.

A community-level composting plant in a rural area in Germany.

Compost is rich in nutrients. It is used in gardens, landscaping, horticulture, and agriculture. The compost itself is beneficial for the land in many ways, including as a soil conditioner, a fertilizer, addition of vital humus or humic acids, and as a natural pesticide for soil. In ecosystems, compost is useful for erosion control, land and stream reclamation, wetland construction, and as landfill cover (see compost uses). Organic ingredients intended for composting can alternatively be used to generate biogas through anaerobic digestion.

TERMINOLOGY

Composting of waste is an aerobic (in the presence of air) method of decomposing solid wastes. The process involves decomposition of organic waste into humus known as compost which is a good fertiliser for plants. However, the term "composting" is used worldwide with differing meanings. Some composting textbooks narrowly define composting as being an aerobic form of decomposition, primarily by microbes. An alternative term to composting is "aerobic digestion", which in turn is also referred to as "wet composting".

For many people, composting is used to refer to several different types of biological process. In North America, "anaerobic composting" is still a common term for what much of the rest of the world and in technical publications people call "anaerobic digestion". The microbes used and the processes involved are quite different between composting and anaerobic digestion.

INGREDIENTS

Home compost barrel in the Escuela Barreales, Santa Cruz, Chile

Carbon, Nitrogen, Oxygen, Water

Materials in a compost pile.

Food scraps compost heap.

Composting organisms require four equally important ingredients to work effectively:

- Carbon — for energy; the microbial oxidation of carbon produces the heat, if included at suggested levels.
- High carbon materials tend to be brown and dry.
- Nitrogen — to grow and reproduce more organisms to oxidize the carbon.
- High nitrogen materials tend to be green (or colorful, such as fruits and vegetables) and wet.
- Oxygen — for oxidizing the carbon, the decomposition process.
- Water — in the right amounts to maintain activity without causing anaerobic conditions.

Certain ratios of these materials will provide beneficial bacteria with the nutrients to work at a rate that will heat up the pile. In that process much water will be released as vapor ("steam"), and the oxygen will be quickly depleted, explaining the need to actively manage the pile. The hotter the pile gets, the more often added air and water is necessary; the air/water balance is critical to maintaining high temperatures (135°-160° Fahrenheit / 50° - 70° Celsius) until the materials are broken down. At the same time, too much air or water also slows the process, as does too much carbon (or too little nitrogen). Hot container composting focuses on retaining the heat to increase decomposition rate and produce compost quicker.

The most efficient composting occurs with an optimal carbon:nitrogen ratio of about 10:1 to 20:1. Rapid composting is favored by having a C/N ratio of ~30 or less. Theoretical analysis is confirmed by field tests that above 30 the substrate is nitrogen starved, below 15 it is likely to outgas a portion of nitrogen as ammonia. If nitrogen needs to be increased, it has been suggested to add 0.15 pounds of *actual* nitrogen per three bushels (3.75 cubic feet) of lower nitrogen material. [For those not familiar with these types of units: 0.64g/L or 640 grams of actual nitrogen per cubic meter.] Two to 3 pounds of organic nitrogen supplement (blood meal, manure, bone meal, alfalfa meal) per 100 pounds of low nitrogen materials (for example, straw or sawdust), supplies generally ample nitrogen and trace minerals in high carbon mixes.

Nearly all plant and animal materials have both carbon and nitrogen, but amounts vary widely, with characteristics noted above (dry/wet, brown/green). Fresh grass clippings have an average ratio of about 15:1 and dry autumn leaves about 50:1 depending on species. Mixing equal parts by volume approximates the ideal C:N range. Few individual situations will provide the ideal mix of materials at any point. Observation of amounts, and consideration of different materials as a pile is built over time, can quickly achieve a workable technique for the individual situation.

Animal Manure and Bedding

On many farms, the basic composting ingredients are animal manure generated on the farm and bedding. Straw and sawdust are common bedding materials. Non-traditional bedding materials are also used, including newspaper and chopped cardboard. The amount of manure composted on a livestock farm is often determined by cleaning schedules, land availability, and weather conditions. Each type of manure has its own physical, chemical, and biological characteristics. Cattle and horse manures, when mixed with bedding, possess good qualities for composting. Swine manure, which is very wet and usually not mixed with bedding material, must be mixed with straw or similar raw materials. Poultry manure also must be blended with carbonaceous materials - those low in nitrogen preferred, such as sawdust or straw.

Microorganisms

With the proper mixture of water, oxygen, carbon, and nitrogen, micro-organisms are allowed to break down organic matter to produce compost. The composting process is dependent on micro-organisms to break down organic matter into compost. There are many types of microorganisms found in active compost of which the most common are:

- Bacteria- The most numerous of all the microorganisms found in compost. Depending on the phase of composting, mesophilic or thermophilic bacteria may predominate.
- Actinobacteria- Necessary for breaking down paper products such as newspaper, bark, etc.
- Fungi- Molds and yeast help break down materials that bacteria cannot, especially lignin in woody material.
- Protozoa- Help consume bacteria, fungi and micro organic particulates.
- Rotifers- Rotifers help control populations of bacteria and small protozoans.

In addition, earthworms not only ingest partly composted material, but also continually re-create aeration and drainage tunnels as they move through the compost.

A lack of a healthy micro-organism community is the main reason why composting processes are slow in landfills with environmental factors such as lack of oxygen, nutrients or water being the cause of the depleted biological community.

Phases of Composting

Under ideal conditions, composting proceeds through three major phases:

- An initial, mesophilic phase, in which the decomposition is carried out under moderate temperatures by mesophilic microorganisms.
- As the temperature rises, a second, thermophilic phase starts, in which the decomposition is carried out by various thermophilic bacteria under high temperatures.
- As the supply of high-energy compounds dwindles, the temperature starts to decrease, and the mesophiles once again predominate in the maturation phase.

Human Waste

Human waste (excreta) can also be added as an input to the composting process, like it is done in composting toilets, as human waste is a nitrogen-rich organic material.

People excrete far more water-soluble plant nutrients (nitrogen, phosphorus, potassium) in urine than in feces. Human urine can be used directly as fertilizer or it can be put onto compost. Adding a healthy person's urine to compost usually will increase temperatures and therefore increase its ability to destroy pathogens and unwanted seeds. Urine from a person with no obvious symptoms of infection is much more sanitary than fresh feces. Unlike feces, urine does not attract disease-spreading flies (such as house flies or blow flies), and it does not contain the most hardy of pathogens, such as parasitic worm eggs. Urine usually does not stink for long, particularly when it is fresh, diluted, or put on sorbents.

Urine is primarily composed of water and urea. Although metabolites of urea are nitrogen fertilizers, it is easy to over-fertilize with urine, or to utilize urine containing pharmaceutical (or other) content, creating too much ammonia for plants to absorb, acidic conditions, or other phytotoxicity.

Humanure

"Humanure" is a portmanteau of *human* and *manure*, designating human excrement (feces and urine) that is recycled via composting for agricultural or other purposes. The term was first used in a 1994 book by Joseph Jenkins that advocates the use of this organic soil amendment. The term humanure is used by compost enthusiasts in the US but not generally elsewhere. Because the term "humanure" has no authoritative definition it is subject to various uses; news reporters occasionally fail to correctly distinguish between humanure and sewage sludge or "biosolids".

USES

Compost is generally recommended as an additive to soil, or other matrices such as coir and peat, as a tilth improver, supplying humus and nutrients. It provides a rich *growing medium*, or a porous, absorbent material that holds moisture and soluble minerals, providing the support and nutrients in which plants can flourish, although it is rarely used alone, being primarily mixed with soil, sand, grit, bark chips, vermiculite, perlite, or clay granules to produce loam. Compost can be tilled directly into the soil or growing medium to boost the level of organic matter and the overall fertility of the soil. Compost that is ready to be used as an additive is dark brown or even black with an earthy smell.

Generally, direct seeding into a compost is not recommended due to the speed with which it may dry and the possible presence of phytotoxins that may inhibit germination, and the possible tie up of nitrogen by incompletely decomposed lignin. It is very common to see blends of 20–30% compost used for transplanting seedlings at cotyledon stage or later.

Composting can destroy pathogens or unwanted seeds. Unwanted living plants (or weeds) can be discouraged by covering with mulch/compost. The microbial pesticides in compost may include thermophiles and mesophiles, however certain composting detritivores such as black soldier fly larvae and redworms, also reduce many pathogens. Thermophilic (high-temperature) composting is well known to destroy many seeds and nearly all types of pathogens (exceptions may include prions). The sanitizing qualities of (thermophilic) composting are desirable where there is a high likelihood of pathogens, such as with manure.

COMPOSTING TECHNOLOGIES

A homemade compost tumbler.

A modern compost bin constructed from plastics.

Overview

In addition to the traditional compost pile, various approaches have been developed to handle different composting processes, ingredients, locations, and applications for the composted product.

There are a large number of different composting systems on the market, for example:

- At the household level: Composting toilet, container composting, vermicomposting
- At the industrial composting (large scale): Aerated Static Pile Composting, vermicomposting, windrow composting etc.

Examples

Vermicomposting

Rotary screen harvested worm castings.

Food waste - after three years.

Vermicompost is the product or process of composting through the utilization of various species of worms, usually red wigglers, white worms, and earthworms, to create a heterogeneous mixture of decomposing vegetable or food waste (excluding meat, dairy, fats, or oils), bedding materials, and vermicast. Vermicast, also known as worm castings, worm humus or worm manure, is the end-product of the breakdown of organic matter by species of earthworm. Vermicomposting is widely used in North America for on-site institutional processing of food waste, such as in hospitals and shopping malls. This type of composting is sometimes suggested as a feasible indoor home composting method. Vermicomposting has gained popularity in both these industrial and domestic settings because, as compared with conventional composting, it provides a way to compost organic materials more quickly (as defined by a higher rate of carbon-to-nitrogen ratio increase) and to attain products that have lower salinity levels that are therefore more beneficial to plant mediums.

The earthworm species (or **composting worms**) most often used are red wigglers (*Eisenia fetida* or *Eisenia andrei*), though European nightcrawlers (*Eisenia hortensis* or *Dendrobaena veneta*) could also be used. Red wigglers are recommended by most vermiculture experts, as they have some of the best appetites and breed very quickly. Users refer to European nightcrawlers by a variety of other names, including *dendrobaenas*, *dendras*, Dutch Nightcrawlers, and Belgian nightcrawlers.

Containing water-soluble nutrients, vermicompost is a nutrient-rich organic fertilizer and soil conditioner in a form that is relatively easy for plants to absorb. Worm castings are sometimes used as an organic fertilizer. Because the earthworms grind and uniformly mix minerals in simple forms, plants need only minimal effort to obtain them. The worms' digestive systems also add beneficial microbes to help create a "living" soil environment for plants.

Vermicompost tea in conjunction with 10% castings has been shown to cause up to a 1.7 times growth in plant mass over plants grown without.

Researchers from the Pondicherry University discovered that worm composts can also be used to clean up heavy metals. The researchers found substantial reductions in heavy metals when the worms were released into the garbage and they are effective at removing lead, zinc, cadmium, copper and manganese.

Hügelkultur (Raised Garden Beds or Mounds)

An almost completed Hügelkultur bed; the bed does not have soil on it yet.

The practice of making raised garden beds or mounds filled with rotting wood is also called "Hügelkultur" in German. It is in effect creating a Nurse log that is covered with soil.

Benefits of hügelkultur garden beds include water retention and warming of soil. Buried wood becomes like a sponge as it decomposes, able to capture water and store it for later use by crops planted on top of the hügelkultur bed.

The buried decomposing wood will also give off heat, as all compost does, for several years. These effects have been used by Sepp Holzer to enable fruit trees to survive at otherwise inhospitable temperatures and altitudes.

Black Soldier fly Larvae Composting

Hermetia illucens § Uses in composting or as food for animals

Black Soldier Fly (*Hermetia illucens*) larvae have been shown to be able to rapidly consume large amounts of organic waste when kept at 31.8 °C, the optimum temperature for reproduction. Enthusiasts have experimented with a large number of different waste products and some even sell starter kits to the public.

Cockroach Composting

Cockroach composting is another insect-mediated composting method. In this case the adults of any number of cockroach species (such as the Turkestan cockroach or *Blaptica dubia*) are used to quickly convert manure or kitchen waste to nutrient dense compost. Depending on species used and environmental conditions, excess composting insects can be used as an excellent animal feed for farm animals and pets.

Bokashi

Inside a recently started bokashi bin. The aerated base is just visible through the food scraps and bokashi bran.

Bokashi is a method that uses a mix of microorganisms to cover food waste or wilted plants to decrease smell. Bokashi (ぼかし) is Japanese for "shading off" or "gradation." It derives from the practice of Japanese farmers centuries ago of covering food waste with rich, local soil that contained the microorganisms that would ferment the waste. After a few weeks, they would bury the waste.

Most practitioners obtain the microorganisms from the product Effective Microorganisms (EM1), first sold in the 1980s. EM1 is mixed with a carbon base (e.g. sawdust or bran) that it sticks to and a sugar for food (e.g. molasses). The mixture is layered with waste in a sealed container and after a few weeks, removed and buried.

Newspaper fermented in a lactobacillus culture can be substituted for bokashi bran for a successful bokashi bucket.

Compost tea

Compost teas are defined as water extracts brewed from composted materials and can be derived from aerobic or anaerobic processes. Compost teas are generally produced from adding one volume of compost to 4-10 volumes of water, but there has also been debate about the benefits of aerating the mixture. Field studies have shown the benefits of adding compost teas to crops due to the adding of organic matter, increased nutrient availability and increased microbial activity. They have also been shown to have an effect on plant pathogens.

Composting Toilets

A composting toilet does not require water or electricity, and when properly managed does not smell. A composting toilet collects human excreta which is then added to a compost heap together with sawdust and straw or other carbon rich materials, where pathogens are destroyed to some extent. The amount of pathogen destruction depends on the temperature (mesophilic or thermophilic conditions) and composting time. A composting toilet tries to process the excreta in situ although this is often coupled with a secondary external composting step. The resulting compost product has been given various names, such as humanure and EcoHumus.

A composting toilet can aid in the conservation of fresh water by avoiding the usage of potable water required by the typical flush toilet. It further prevents the pollution of ground water by controlling the fecal matter decomposition before entering the system. When properly managed, there should be no ground contamination from leachate.

COMPOST AND LAND-FILLING

As concern about landfill space increases, worldwide interest in recycling by means of composting is growing, since composting is a process for converting decomposable organic materials into useful stable products. Composting is one of the only ways to revitalize soil vitality due to phosphorus depletion in soil.Industrial scale composting in the form of in-vessel composting, aerated static pile composting, and anaerobic digestion takes place in most Western countries now, and in many areas is mandated by law. There are process and product guidelines in Europe that date to the early 1980s (Germany, the Netherlands, Switzerland) and only more recently in the UK and the US. In both these countries, private trade associations within the industry have established loose standards, some say as a stop-gap measure to discourage independent government agencies from establishing tougher consumer-friendly standards. The USA is the only Western country that does not distinguish sludge-source compost from green-composts, and by default in the USA 50% of states expect composts to comply in some manner with the federal EPA 503 rule promulgated in 1984 for sludge products. Compost is regulated in Canada and Australia as well.

Industrial Systems

A large compost pile that is steaming with the heat generated by thermophilic microorganisms.

Industrial composting systems are increasingly being installed as a waste management alternative to landfills, along with other advanced waste processing systems. Mechanical sorting of mixed waste streams combined with anaerobic digestion or in-vessel composting is called mechanical biological treatment, and is increasingly being used in developed countries due to regulations controlling the amount of organic matter allowed in landfills. Treating biodegradable waste before it enters a landfill reduces global warming from fugitive methane; untreated waste breaks down anaerobically in a landfill, producing landfill gas that contains methane, a potent greenhouse gas.

Vermicomposting, also known as vermiculture, is used for medium-scale on-site institutional composting, such as for food waste from universities and shopping malls. It is selected either as a more environmentally friendly choice than conventional methods of disposal, or to reduce the cost of commercial waste removal.

Large-scale composting systems are used by many urban areas around the world. Co-composting is a technique that combines solid waste with de-watered biosolids, although difficulties controlling inert and plastics contamination from municipal solid waste makes this approach less attractive. The world's largest MSW co-composter is the Edmonton Composting Facility in Edmonton, Alberta, Canada, which turns 220,000 tonnes of residential solid waste and 22,500 dry tonnes of biosolids per year into 80,000 tonnes of compost. The facility is 38,690 m^2 (416,500 sq.ft.) in area, equivalent to 4½ Canadian football fields, and the operating structure is the largest stainless steel building in North America, the size of 14 NHL rinks. In 2006, Qatar awarded Keppel Seghers Singapore, a subsidiary of Keppel Corporation, a contract to begin construction on a 275,000 tonne/year anaerobic digestion and composting plant licensed by Kompogas (de) Switzerland. This plant, with 15 independent anaerobic digesters, will be the world's largest composting facility once fully operational in early 2011 and forms part of Qatar's Domestic Solid Waste Management Centre, the largest integrated waste management complex in the Middle East.

Another large MSW composter is the Lahore Composting Facility in Lahore, Pakistan, which has a capacity to convert 1,000 tonnes of municipal solid waste per day into compost. It also has a capacity to convert substantial portion of the intake into refuse-derived fuel (RDF) materials for further combustion use in several energy consuming industries across Pakistan, for example in cement manufacturing companies where it is used to heat cement kilns. This project has also been approved by the Executive Board of the United Nations Framework Convention on Climate Change for reducing methane emissions, and has been registered with a capacity of reducing 108,686 tonnes carbon dioxide equivalent per annum.

RELATED TECHNOLOGIES

Anaerobic digestion is process for converting organic waste into (biogas). The residual material, sometimes in combination with sewage sludge can be followed by an aerobic composting process before selling or giving away the compost.

HISTORY

Composting as a recognized practice dates to at least the early Roman Empire since Pliny the Elder (AD 23-79). Traditionally, composting involved piling organic materials until the next planting season, at which time the materials would have decayed enough to be ready for use in the soil. The advantage of this method is that little working time or effort is required from the composter and it fits in naturally with agricultural practices in temperate climates. Disadvantages (from the modern perspective) are that space is used for a whole year, some nutrients might be leached due to exposure to rainfall, and disease-producing organisms and insects may not be adequately controlled.

Composting was somewhat modernized beginning in the 1920s in Europe as a tool for organic farming. The first industrial station for the transformation of urban organic materials into compost was set up in Wels, Austria in the year 1921. Early frequent citations for propounding composting within farming are for the German-speaking world Rudolf Steiner, founder of a farming method called biodynamics, and Annie Francé-Harrar, who was appointed on behalf of the government in Mexico and supported the country 1950–1958 to set up a large humus organization in the fight against erosion and soil degradation.

Compost Basket.

In the English-speaking world it was Sir Albert Howard who worked extensively in India on sustainable practices and Lady Eve Balfour who was a huge proponent of composting. Composting was imported to America by various followers of these early European movements by the likes of J.I. Rodale (founder of Rodale Organic Gardening), E.E. Pfeiffer (who developed scientific practices in biodynamic farming), Paul Keene (founder of Walnut Acres in Pennsylvania), and Scott and Helen Nearing (who inspired the back-to-the-land movement of the 1960s). Coincidentally, some of the above met briefly in India - all were quite influential in the U.S. from the 1960s into the 1980s.

There are many modern proponents of rapid composting that attempt to correct some of the perceived problems associated with traditional, slow composting. Many advocate that compost can be made in 2 to 3 weeks. Many such short processes involve a few changes to traditional methods, including smaller, more homogenized pieces in the compost, controlling carbon-to-nitrogen ratio (C:N) at 30 to 1 or less, and monitoring the moisture level more carefully. However, none of these parameters differ significantly from the early writings of Howard and Balfour, suggesting that in fact modern composting has not made significant advances over the traditional methods that take a few months to work. For this reason and others, many modern scientists who deal with carbon transformations are sceptical that there is a super-charged way to get nature to make compost rapidly.

In fact, both sides are right to some extent. The bacterial activity in rapid high heat methods breaks down the material to the extent that pathogens and seeds are destroyed, and the original feedstock is unrecognizable. At this stage, the compost can be used to prepare fields or other planting areas. However, most professionals recommend that the compost be given time to cure before using in a nursery for starting seeds or growing young plants. The curing time allows fungi to continue the decomposition process and eliminating phytotoxic substances.

Many countries such as Wales and some individual cities such as Seattle and San Francisco require food and yard waste to be sorted for composting.

Kew Gardens in London has one of the biggest non-commercial compost heaps in Europe.

COMPOSTING TOILET

Composting toilet at Activism Festival 2010 in the mountains outside Jerusalem.

A **composting toilet** is a type of dry toilet that uses a predominantly aerobic processing system to treat human excreta, by composting or managed aerobic decomposition. These toilets generally use little to no water and may be used as an alternative to flush toilets. They have found use in situations where no suitable water supply or sewer system and sewage treatment plant is available to capture the nutrients in human excreta. They are in use in many roadside facilities and national parks in Sweden, Canada, US, UK and Australia. They are used in rural holiday homes in Sweden and Finland.

The human excreta is usually mixed with sawdust, coconut coir or peat moss to facilitate aerobic processing, liquid absorption, and odor mitigation. Most composting toilets use slow, cold composting conditions, sometimes connected to a secondary external composting step.

Composting toilets produce a compost that may be used for horticultural or agricultural soil enrichment if the local regulations allow this. A curing stage is often needed to allow mesophilic composting to reduce potential phytotoxins.

TERMINOLOGY

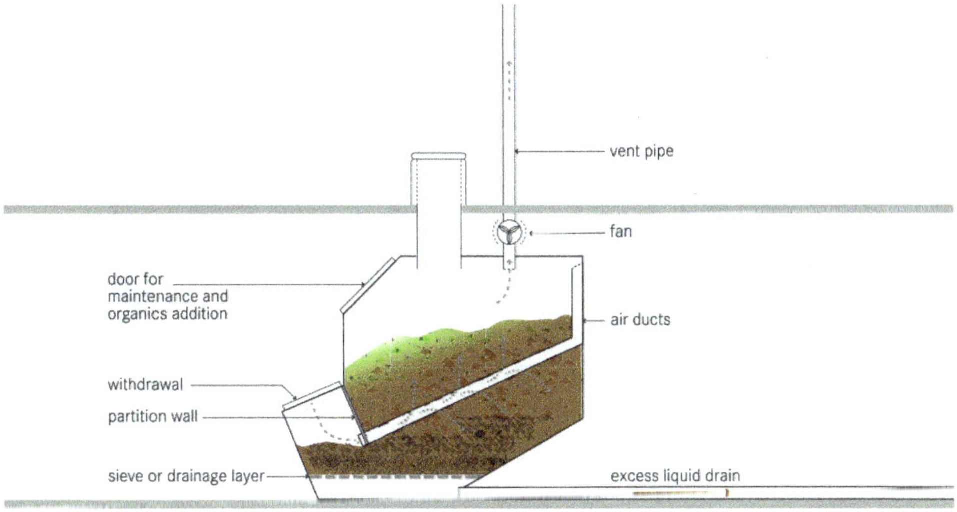

Schematic of the composting chamber which is located below the toilet seat

The term "composting toilet" is used quite loosely, and its meaning may vary by country. For example, in English-speaking countries, the term "anaerobic composting" (equivalent to anaerobic decomposition) is used. In Germany and Scandinavian countries, composting always refers to a predominantly aerobic process. This aerobic composting may take place with an increase in temperature due to microbial action, or without a temperature increase in the case of slow composting or cold composting. If earth worms are used (vermicomposting) then there is also no increase in temperature.

Composting toilets differ from pit latrines, arborloo or tree bogs, all of which are forms of less controlled decomposition and may not protect groundwater from nutrient or pathogen contamination or provide optimal nutrient recycling. They also differ from urine-diverting dry toilets (UDDTs) where pathogen reduction is achieved through dehydration (also known by the more precise term "desiccation) and where the faeces collection vault is kept as dry as possible. Composting toilets target a certain degree of moisture in the composting chamber.

Composting toilets usually do not divert urine. Offering a waterless urinal in addition to the toilet can help keep excess amounts of urine out of the composting chamber.

Composting toilets can be used to implement an ecological sanitation approach for resource recovery, and some people call their composting toilet designs ecosan toilets for that reason. However, this is not recommended as the two terms (i.e. composting and ecosan) are not identical.

Composting toilets have also been called sawdust toilets, which can be appropriate if the amount of aerobic composting taking place in the toilet's container is very limited.

The Clivus multrum is a type of composting toilet which has a large composting chamber below the toilet seat and also receives undigested organic material to increase the carbon to nitrogen ratio.

Applications

This is the pedestal for a split-system composting toilet where collection/treatment chambers are located below the bathroom floor.

Inexpensive do-it-yourself compost toilet at Dial House, Essex, England, utilizing an old desk as the toilet unit.

Public composting toilet at a highway rest facility in Sweden.

Composting toilets can be suitable in areas such as a rural area or a park that lacks a suitable water supply, sewers and sewage treatment. They can also help increase the resilience of existing sanitation systems in the face of possible natural disasters such as climate change, earthquakes or tsunami. Composting toilets can reduce or perhaps eliminate the need for a septic tank system to reduce environmental footprint (particularly when used in conjunction with an on-site greywater treatment system).

These types of toilets can be used for resource recovery by reusing sanitized feces and urine as fertilizer and soil conditioner for gardening or ornamental activities.

Basics

Components

A composting toilet consists of two elements: a place to sit or squat and a collection/ composting unit. The composting unit consists of four main parts:

- storage or composting chamber
- a ventilation unit to ensure that the degradation process in the toilet is predominantly aerobic and to vent odorous gases
- a leachate collection system to remove excess liquid
- an access door for extracting the compost

Construction

The composting chamber can be constructed above or below ground level. It can be inside a structure or include a separate superstructure.

A drainage system removes leachate. Otherwise, excess moisture can cause anaerobic conditions and impede decomposition. Urine diversion can improve compost quality, since urine contains large amounts of ammonia that inhibits microbiological activity.

Composting toilets greatly reduce human waste volumes through psychrophilic, thermophilic or mesophilic composting. Keeping the composting chamber insulated and warm protects the composting process from slowing due to low temperatures.

Odorous Gases

The following gases may be emitted during the composting process that takes place in composting toilets: hydrogen sulfide (H_2S), ammonia, nitrous oxide (N_2O) and volatile organic compounds (VOCs). These gases can potentially lead to complaints about odours. Some methane may also be present, but it is not odorous.

Pathogen Removal

Excreta-derived compost recycles fecal nutrients, but it can carry and spread pathogens if the process of reuse of excreta is not done properly.

Internal pathogen destruction rates are usually low, particularly helminth eggs, such as Ascaris eggs. This carries the risk of spreading disease if a proper system management is not in place. Compost from human excreta processed under only mesophilic conditions or taken directly from the compost chamber is not safe for food production. High temperatures or long composting times are required to kill helminth eggs, the hardiest of all pathogens. Helminth infections are common in many developing countries.

In thermophilic composting bacteria that thrive at temperatures of 40–60 °C (104–140 °F) oxidize (break down) waste into its components, some of which are consumed in the process, reducing volume and eliminating potential pathogens. To destroy pathogens, thermophilic composting must heat the compost pile sufficiently, or enough time (1–2 years) must elapse since fresh material was added that biological activity has had the same pathogen removal effect.

One guideline claims that pathogen levels are reduced to a safe level by thermophilic composting at temperatures of 55 °C for at least two weeks or at 60 °C for one week. An alternative guideline claims that complete pathogen destruction may be achieved already if the entire compost heap reaches a temperature of 62 °C (144 °F) for one hour, 50 °C (122 °F) for one day, 46 °C (115 °F) for one week or 43 °C (109 °F) for one month, although others regard this as overly optimistic.

Design Considerations

Composting toilet with a seal in the lid in Germany.

Environmental Factors

Four main factors affect the decomposition process:

- Sufficient oxygen is necessary for aerobic composting
- Moisture content from 45 to 70 percent (heuristically, "the compost should feel damp to the touch, with only a drop or two of water expelled when tightly squeezed in the hand.")
- Temperature between 40 and 50 °C (achieved through proper chamber dimensioning and possibly active mixing)
- Carbon-to-nitrogen ratio (C:N) of 25:1

Additives and Bulking Material

Human excreta and food waste do not provide optimum conditions for composting. Usually the water and nitrogen content is too high, particularly when urine is mixed with feces. Additives or "bulking material", such as wood chips, bark chips, sawdust, ash and pieces of paper can absorb moisture. The additives improve pile aeration and increase the carbon to nitrogen ratio. Bulking material also covers faeces and reduces insect access. Absent sufficient bulking material, the material may become too compact and form impermeable layers, which leads to anaerobic conditions and odour.

Leachate management

Leachate removal controls moisture levels, which is necessary to ensure rapid, aerobic composting. Some commercial units include a urine-separator or urine-diverting system and/or a drain at the bottom of the composter for this purpose.

Aeration and mixing

Microbial action also requires oxygen, typically from the air. Commercial systems provide ventilation that moves air from the bathroom, through the waste container, and out a vertical pipe, venting above the roof. This air movement (via convection or fan forced) passes carbon dioxide and odors.

Some units require manual methods for periodic aeration of the solid mass such as rotating the composting chamber or pulling an aerator rake through the mass.

Types

External composting chamber of a composting toilet at a house in France.

Commercial units and construct-it-yourself systems are available. Variations include number of composting vaults, removable vault, urine diversion and active mixing/aeration.

Slow Composting (or moldering) Toilets

Most composting toilets use slow composting which is also called “cold composting”. The compost heap is built up step by step over time.

The finished end product from "slow" composting toilets ("moldering toilets" or "moldering privies" in the US), is generally not free of pathogens. World Health Organization Guidelines from 2006 offer a framework for safe reuse of excreta, using a multiple barrier approach.

Slow composting toilets employ a passive approach. Common applications involve modest and often seasonal use, such as remote trail networks. They are typically designed such that the materials deposited can be isolated from the operational part. The toilet can also be closed to allow further mesophilic composting. Slow composting toilets rely on long retention times for pathogen reduction and for decomposition of excreta or on the combination of time and/or the addition of red wriggler worms for vermi-composting. Worms can be introduced to accelerate composting. Some jurisdictions of the US consider these worms as invasive species.

Example in Vermont Woods

Slow composting toilets have been installed by the Green Mountain Club in Vermont's woodlands. They employ multiple vaults (called cribs) and a movable building. When one of the vaults fills, the building is moved over an empty vault. The full vault is left untouched for as long as possible (up to three years) before it is emptied. The large surface area and exposure to air currents can cause the pile to dry out. To counteract this, signs instruct users to urinate in the toilet. The club also uses pit latrines and simple bucket toilets with woodchips and external composting and directs users to urinate in the forest to prevent odiferous anaerobic conditions.

Active Composters

Self-Contained

"Self-contained" composting toilets compost in a container within the toilet unit. They are slightly larger than a flush toilet, but use roughly the same floor space. Some units use fans for aeration, and optionally, heating elements to maintain optimum temperatures to hasten the composting process and to evaporate urine and other moisture. Operators of composting toilets commonly add a small amount of absorbent carbon material (such as untreated sawdust, coconut coir, peat moss) after each use to create air pockets to encourage aerobic processing, to absorb liquid and to create an odor barrier. This additive is sometimes referred to as "bulking agent." Some owner-operators use microbial "starter" cultures to ensure composting bacteria are in the process, although this is not critical.

Remote

"Remote" "central" or "underfloor" units collect excreta via a toilet stool, either waterless, vacuum or micro-flush, from which it drains into a composter. "Vacuum-flush systems" can flush horizontally or upward with a small amount of water to the composter. "Micro-flush" toilets use about 500 millilitres (17 US fl oz) per use. These units feature a chamber below the toilet stool (such as in a basement or outside) where composting takes place and are suitable for high-volume and year-round applications as well as to serve multiple toilet stools.

Other

Some units employ roll-away containers fitted with aerators, while others use sloped-bottom tanks.

Maintenance

Maintenance is critical to ensure proper operation, including odor prevention. Maintenance tasks include: cleaning, servicing technical components such as fans and removal of compost, leachate and urine. Urine removal is only required for those types of composting toilets using urine diversion.

Once composting is complete (or more often), the compost must be removed from the unit. How often this occurs is a function of container size, usage and composting conditions, such as temperature. Active, hot composting may span months only while passive, cold composting may require years. Properly managed units yield output volumes of about 10% of inputs.

Uses of Compost

Finished compost from a composting toilet ready for application as soil improvement in Kiel-Hassee, Germany

The material from composting toilets is a humus-like material, which can be suitable as a soil amendment for agriculture. Compost from residential composting toilets can be used in domestic gardens, and this is the main such use.

Enriching soil with compost adds substantial nitrogen, phosphorus, potassium, carbon and calcium. In this regard compost is equivalent to many fertilizers and manures purchased in garden stores. Compost from composting toilets has a higher nutrient availability than the dried faeces that result from a urine-diverting dry toilet.

Urine is typically present, although some is lost via leaching and evaporation. Urine can contain up to 90 percent of the residual nitrogen, up to 50 percent of the phosphorus, and up to 70 percent of the potassium.

Compost derived from these toilets has in principle the same uses as compost derived from other organic waste products, such as sewage sludge or municipal organic waste. However, users of excreta-derived compost must consider the risk of pathogens.

Pharmaceutical Residues

Excreta-derived compost may contain prescription pharmaceuticals. Such residues are also present in conventional wastewater treatment effluent. This could contaminate groundwater. Among the medications that have been found in groundwater in recent years are antibiotics, antidepressants, blood thinners, ACE inhibitors, calcium-channel blockers, digoxin, estrogen, progesterone, testosterone, Ibuprofen, caffeine, carbamazepine, fibrates and cholesterol-reducing medications. Between 30% and 95% of pharmaceuticals medications are excreted by the human body. Medications that are lipophilic (dissolved in fats) are more likely to reach groundwater by leaching from fecal wastes. Wastewater treatment plants remove an average of 60% of these medications. The percentage of medications degraded during composting of excreta has not yet been reported.

Comparison

Pit Latrines

Unlike pit latrines, composting toilets convert feces into a dry, odorless material, avoiding the issues surrounding liquid fecal sludge management (e.g. odor, insects and disposal). These toilets minimize the risk of water pollution through the safe containment of feces in above-ground vaults, which allows the toilets to be sited in locations where pit-based systems are not appropriate. However, composting toilets face higher capital costs (although lifecycle costs might be lower) and greater complexity (for instance, adding covering materials and managing moisture content).

Flush Toilets

Unlike flush toilets, composting toilets do not dilute excreta and create wastewater streams which must be treated before disposal. On the other hand, wastewater treatment plants can centralize waste management for an entire community, with potentially greater efficiency.

Urine-Diverting Dry Toilets

Composting toilets are more difficult to maintain than other types of dry toilets, like urine-diverting dry toilets (UDDT) with which they are often confused. This is due to the need to maintain a consistent and relatively high moisture content, as well as the relatively high complexity of composting toilets compared to UDDTs. Apart from that, composting toilets are quite similar to UDDTs, sharing many of the same advantages and disadvantages.

Henry Moule's earth closet, patented in 1873 (not a true composting toilet). Example from around 1875. Rear chamber for dispensing cover material.

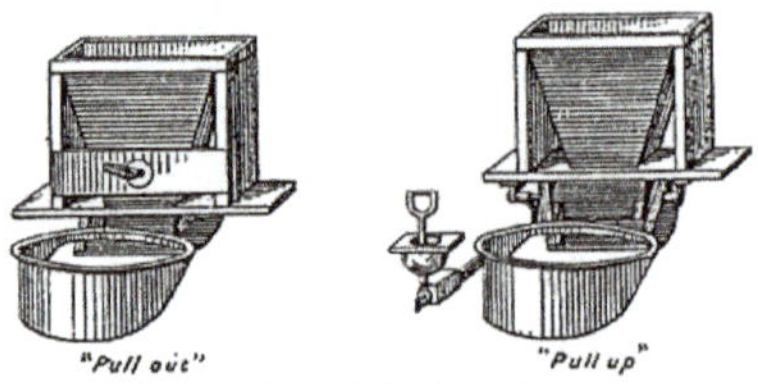

FIGS. 8 and 9.—Moule's Earth Closets.

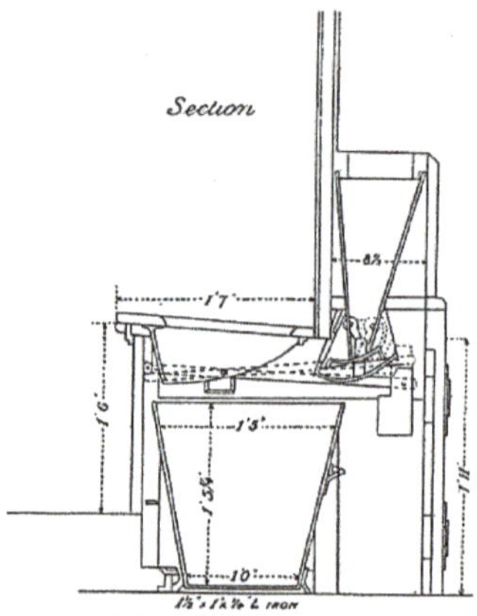

FIG. 10.—Section of Moule's Automatic Earth Closet.

Henry Moule's earth closet design, circa 1909.

Dry Earth Toilet

Before the flush toilet became accepted in the late 19th century in developed countries, some inventors, scientists and public health officials supported the use of "dry earth closets", a type of dry toilet with similarities to composting toilets, but the collection vessel for the human excreta was not designed to compost. Dry earth closets were invented by English clergyman Henry Moule, who dedicated his life to improving public sanitation after witnessing the cholera epidemics of 1849 and 1854. Impressed by the insalubrity of the houses, especially during the Great Stink in the summer of 1858, he invented what he called the 'dry earth system'.

In partnership with James Bannehr, he patented his device (No. 1316, dated 28 May 1860). Among his works bearing on the subject were *The Advantages of the Dry Earth System* (1868), *The Impossibility overcome: or the Inoffensive, Safe, and Economical Disposal of the Refuse of Towns and Villages (*1870), The *Dry Earth System* (1871), *Town Refuse, the Remedy for Local Taxation* (1872), and *National Health and Wealth promoted by the general adoption of the Dry Earth System* (1873).

His system was adopted in private houses, in rural districts, in military camps, in many hospitals, and extensively in the British Raj. Ultimately, however, it failed to gain public support as attention turned to the water-flushed toilet connected to a sewer system.

In Germany, a similar dry toilet with a peat dispenser was marketed until after the second World War (it was called Metroclo and was manufactured by Gefinal, Berlin).

Society and Culture

Regulations

International Organization for Standardization (ISO)

The International Organization for Standardization (ISO) is currently preparing a management standard. As of 2015 this was in a draft state as ISO 24521, under the heading Activities relating to drinking water and wastewater services — Guidelines for the management of basic onsite domestic wastewater services. The standard is meant to be used in conjunction with ISO 24511. It deals with toilets (including composting toilets) and toilet waste. The guidelines are applicable to basic wastewater systems and include the complete domestic wastewater cycle, such as planning, usability, operation and maintenance, disposal, reuse and health.

International Association of Plumbing and Mechanical Officials

The International Association of Plumbing and Mechanical Officials (IAPMO) is a plumbing and mechanical code structure adopted by many developed countries. It recently proposed an addition to its Green Plumbing Mechanical Code Supplement that, ...outlines performance criteria for site built composting toilets with and without urine diversion and manufactured composting toilets. If adopted, this composting and urine diversion toilet code (the first of its kind in the United States) will appear in the 2015 edition of the Green Supplement to the Uniform Plumbing Code.

United States

No performance standards for composting toilets are universally accepted in the US. Seven jurisdictions in North Americause *American National Standard/NSF International Standard ANSI/NSF 41-1998: Non-Liquid Saturated Treatment Systems*. An updated version was published in 2011.[note 1] Systems might also be listed with the Canadian Standards Association, cETL-US, and other standards programs.

Regarding byproduct regulation, several US states permit disposal of solids from composting toilets (usually a distinction between different types of dry toilets is not made) by burial, with varying or no minimum depth mandates (as little as 6 inches). For instance:

- Massachusetts: Residuals from the composting toilet system must be buried on-site and covered with a minimum of six inches of clean compacted soil. Massachusetts requires that any liquids produced but, not recycled through the toilet [itself be] either discharged through a greywater system on the property that includes a septic tank and soil absorption system, or removed by a licensed septage hauler.
- Oregon: Humus from composting toilets may be used around ornamental shrubs, flowers, trees, or fruit trees and shall be buried under at least twelve inches of soil cover.
- Rhode Island: Solids produced by alternative toilets may be buried on site, while, residuals shall not be applied to food crops.
- Virginia: All materials removed from a composting privy shall be buried, and compost material shall not be placed in vegetable gardens or on the ground surface.
- Vermont: Byproducts may be disposed via ...shallow burial in a location approved by the Agency that meets the minimum site conditions [required for an onsite septic tank-based sanitation system].
- Washington: models its extensive regulations for what it refers to as waterless toilets on the federal regulations that govern sewage sludge.

The Environmental Protection Agency has no jurisdiction over the byproducts of a dry toilet as long as excreta are not referred to as fertilizer (but instead simply a material that is being disposed of). Federal rule 503, known colloquially as the EPA Biosolids rule or the EPA sludge rule applies only to fertilizer. Thus, individual states regulate composting toilets.

Germany

The regulations for composting toilets and other forms of dry toilets in Germany vary from state to state and from one application to another (e.g. use in allotment gardens or use in family homes and settlements). In the different states of Germany, it is the "Landesbauordnung" (translates to "state civil engineering regulations") of the respective state that regulates the use of such alternative toilets. Most of them stipulate the use of flush toilets, however there are many exceptions, for example in the states of Hamburg, Lower Saxony, Bavaria, Mecklenburg-Western Pomerania, Rhineland-Palatinate, Saxony-Anhalt and Thuringia. These generally make exceptions for the use of composting toilets in homes provided that there are no concerns for public health.

Regulations governing the use of compost and urine from composting toilets is less clear in Germany but it seems generally allowed provided it is used on one's own property and not sold to third parties.

Examples

Finland

Numerous sparsely settled villages in rural areas in Finland are not connected to municipal water supply or sewer networks, requiring homeowners to operate their own systems. Individual private wells, i.e. shallow dug wells or boreholes in the bedrock, are often used for water supply, and many homeowners have opted for composting toilets. In addition, these toilets are common at holiday homes, often located near sensitive water bodies. For these reasons, many manufacturers of composting toilets are based in Finland, including Biolan, Ekolet, Kekkilä, Pikkuvihreä and Raita Environment.

Estimates made by leading Finnish composting toilet manufacturers and the Global Dry Toilet Association of Finland provided the following 2014 figures for composting toilet use in Finland:

- About 4% of single-family homes not connected to a public sewer network are equipped with a composting toilet.
- Some 200,000 manufactured composting toilets are thought to serve holiday homes, matched by the number of other dry toilets. The simplest ones are sited in an outhouse.

Germany

Composting container of "TerraNova" composting toilet, showing open removal chamber (town house at the ecological settlement Hamburg-Allermöhe, Germany).

Composting toilets have been successfully installed in houses with up to four floors. An estimate from 2008 put the number of composting toilets in households in Germany at 500. Most of these residences are also connected to a sewer system; the composting toilet was not installed due a lack of sewer system but for other reasons, mainly because of an "ecological mindset" of the owners.

In Germany and Austria, composting toilets and other types of dry toilets have been installed in single and multi-family houses (e.g. Hamburg, Freiburg, Berlin), ecological settlements (e.g. Hamburg-Allcmöhe, Hamburg-Braamwisch, Kiel-Hassee, Bielefeld-Waldquelle, Wien-Gänserndorf) and in public buildings (e.g. Ökohaus Rostock, VHS-Ökostation Stuttgart-Wartberg, public toilets in recreational areas, restaurants and huts in the Alps, house boats and forest Kindergartens).

The ecological settlement in Hamburg-Allermöhe has had composting toilets since 1982. The settlement of 36 single-family houses with approximately 140 inhabitants uses composting toilets, rainwater harvesting and constructed wetlands. Composting toilets save about 40 litres of water per capita per day compared to a conventional flush toilet (10 liter per flush), which adds up to 2,044 m³ water savings per year for the whole settlement.

Worldwide

Composting toilets with a large composting container (of the type Clivus Multrum and derivations of it) are popular in US, Canada, Australia, New Zealand and Sweden. They can be bought and installed as commercial products, as designs for self builders or as "design derivatives" which are marketed under various names. It has been estimated that approximately 10,000 such toilets might be in use worldwide.

REFERENCES

1. Tilley, E.; Ulrich, L.; Lüthi, C.; Reymond, Ph.; Zurbrügg, C. Compendium of Sanitation Systems and Technologies - (2nd Revised Edition). Swiss Federal Institute of Aquatic Science and Technology (Eawag), Duebendorf, Switzerland. p. 72. ISBN 978-3-906484-57-0.
2. Berger, W. (2011). Technology review of composting toilets - Basic overview of composting toilets (with or without urine diversion). Deutsche Gesellschaft für Internationale Zusammenarbeit (GIZ) GmbH, Eschborn, Germany
3. Rieck, C., von Münch, E., Hoffmann, H. (2012). Technology review of urine-diverting dry toilets (UDDTs) - Overview on design, management, maintenance and costs. Deutsche Gesellschaft fuer Internationale Zusammenarbeit (GIZ) GmbH, Eschborn, Germany
4. Hill, B. G. (2013). An evaluation of waterless human waste management systems at North American public remote sites. PhD thesis, University of British Columbia (Vancouver), Canada
5. The online Compendium of Sanitation Systems and Technologies. The online Compendium of Sanitation Systems and Technologies. eawag aquatic research. 2014. Retrieved 2014-12-29.

6. Font, Xavier; Artola, Adriana; Sánchez, Antoni (6 April 2011). Detection, Composition and Treatment of Volatile Organic Compounds from Waste Treatment Plants. Sensors. **11** *(12): 4043–4059. doi:10.3390/s110404043.*

7. Stenström, T.A., Seidu, R., Ekane, N., Zurbrügg, C. (2011). Microbial exposure and health assessments in sanitation technologies and systems - EcoSanRes Series, 2011-1. Stockholm Environment Institute (SEI), Stockholm, Sweden, page 88

8. National Small Flows Clearinghouse, West Virginia University, Composting toilet technology

9. WHO (2006). WHO Guidelines for the Safe Use of Wastewater, Excreta and Greywater - Volume IV: Excreta and greywater use in agriculture. World Health Organization (WHO), Geneva, Switzerland

10. Appalachian Trail Conservancy (2014). Backcountry Sanitation Manual, 2nd Edition. Appalachian Trail Conservancy, Green Mountain Club, USDA Forest Service, National Park Service, USA

11. Allen, Lee (2013). Long Trail News: Quarterly of the Green Mountain Club, Fall 2013. Article titled: A Privy is a Privy is a Privy...or is it? To Pee or Not Pee.(PDF). Green Mountain Club. Green Mountain Club. Retrieved 31 January 2013.

12. Antos-Ketcham, Pete (2013). Long Trail News: Quarterly of the Green Mountain Club, Fall 2013. Article titled: Batch-Bin/Beyond-the-Bin (BTB) Composting Privies (PDF). Green Mountain Club. Green Mountain Club. Retrieved 31 January 2015.

13. Berger, W. (2009). Appendix of technology review of composting toilets - List of manufacturers and commercially available composting toilets. Gesellschaft für Internationale Zusammenarbeit (GIZ) GmbH

14. J.O. Drangert, Urine separation systems

15. Drugs in the Water. Harvard Health Letter. 2011.

16. Encyclopedia of Quantitative Risk Analysis and Assessment, Volume 1, edited by Edward L. Melnick, Brian S. Veritt, 2008

17. ISO/DIS 24521. Activities relating to drinking water and wastewater services -- Guidelines for the management of basic onsite domestic wastewater services. International Organization for Standardization (ISO). Retrieved 15 January 2015.

18. ISO 24511:2007. Activities relating to drinking water and wastewater services -- Guidelines for the management of wastewater utilities and for the assessment of wastewater services. International Organization for Standardization (ISO). Retrieved 15 January 2015.

19. Recode September 2014 Newsletter. Recode. Recode. September 2014. Retrieved 15 January 2015.

20. IAPMO Proposed Composting and Urine DIversion Toilet Code (PDF). The IAPMO Group. International Association of Plumbing and Mechanical Officials. Retrieved 15 January 2015.

21. Cole, Daniel (January 2015). IAPMO GPMCS raising the bar for water, energy efficiency. Plumbing Engineer. Plumbing Engineer. Retrieved 15 January 2015.
22. Oregon Onsite Advisory Committee "Final Report of Recommended Changes to Rules Governing Onsite Systems", OR DEQ, February 8, 2010, accessed May 8, 2011.
23. PUBLICATIONS - Standards and Criteria - March 21, 2013 (PDF). NSF International. p. 4. Retrieved 24 March 2013. Wastewater Treatment Units ... NSF/ANSI 41 – 2011: Non-liquid saturated treatment systems (composting toilets)
24. Regulatory Provisions for Composting Toilets and Greywater Systems. The Official Website of the Massachusetts Executive Office of Energy and Environmental Affairs. Office of Energy and Environmental Affairs. Retrieved 13 January 2015.
25. Department of Consumer and Business Services, Building Codes Division, Division 770, Plumbing Product Approvals. Oregon Secretary of State. State of Oregon. Retrieved 13 January 2015.
26. State of Rhode Island and Providence Plantations Department of Environmental Management, Office of Water Resources: Rules Establishing Minimum Standards Relating to Location, Design, Construction and Maintenance of Onsite Wastewater Treatment Systems (PDF). State of Rhode Island Department of Environmental Management. STATE OF RHODE ISLAND AND PROVIDENCE PLANTATIONS. July 2010. Retrieved 13 January 2015.
27. SEWAGE HANDLING AND DISPOSAL REGULATIONS (Emergency Regulations for Gravelless Material and Drip Dispersal), 12 VAC 5-610-10 et seq. (PDF). State of Virginia Department of Health. Commonwealth of Virginia. 14 March 2014. Retrieved 13 January 2015.
28. Environmental Protection Rules, Chapter 1: Wastewater System and Potable Water Supply Rules (PDF). State of Vermont Drinking Water and Groundwater Protection Division. State of Vermont. 29 September 2007. Retrieved 14 January 2015.
29. Recommended Standards and Guidance for Performance, Application, Design, and Operation & Maintenance: Water Conserving On-Site Wastewater Treatment Systems (PDF). State of Washington Department of Health. State of Washington. July 2012. Retrieved 14 January 2015.
30. Water Efficiency Technology Fact Sheet: Composting Toilets (PDF). United States Environmental Protection Agency, Office of Water, Washington, D.C., EPA 832-F-99-066. United States Environmental Protection Agency, Office of Water. September 1999. Retrieved 13 January 2015.
31. TITLE 40—Protection of Environment, Chapter I—Environmental Protection Agency (Continued), Subchapter O—Sewage Sludge, Part 503—Standards for the Use or Disposal of Sewage Sludge. Electronic Code of Federal Regulations. United States Government Publishing Office. Retrieved 13 January 2015.
32. Lorenz-Ladener, Hrsg. Claudia; Berger, Wolfgang (2005). Kompost-Toiletten: Wege zur sinnvollen Fäkalienentsorgung (1. überarb. u. erw. Aufl. ed.). Staufen im Breisgau: Ökobuch. p. 178. ISBN 978-3-936896-16-9.

33. Global Dry Toilet Association of Finland (2011) Dry Toilet Manufacturers in Finland, Leaflet in English and Finnish

34. Global Dry Toilet Association of Finland. Global Dry Toilet Association of Finland - Company and association members. Retrieved 15 January 2015.

35. Lorenz-Ladener, Hrsg. Claudia; Berger, Wolfgang (2005). Kompost-Toiletten: Wege zur sinnvollen Fäkalienentsorgung (1. überarb. u. erw. Aufl. ed.). Staufen im Breisgau: Ökobuch. p. 183. ISBN 978-3-936896-16-9.

36. Rauschning, G., Berger, W., Ebeling, B., Schöpe, A. (2009). Ecological settlement in Allermöhe Hamburg, Germany - Case study of sustainable sanitation projects. Sustainable Sanitation Alliance (SuSanA)

CHAPTER 5

Aerobic Digestion

Chapter Contents

- **Process**
- **Advantages**
- **Disadvantages**
- **Aerobic Granulation**
- **Aerobic Granular Reactor**
- **Bioreactor**

Aerobic digestion is a process in sewage treatment designed to reduce the volume of sewage sludge and make it suitable for subsequent use. It is a bacterial process occurring in the presence of oxygen. Bacteria rapidly consume organic matter and convert it into carbon dioxide, water and a range of lower molecular weight organic compounds. As there is no new supply of organic material from sewage, the activated sludge biota begin to die and are used as food by saporotrophic bacteria. This stage of the process is known as *endogenous respiration* and it is process that reduces the solid concentration in the sludge.

PROCESS

Aerobic digestion is typically used in an activated sludge treatment plant. Waste activated sludge and primary sludge are combined, where appropriate, and passed to a thickener where the solids content is increased. This substantially reduces the volume that is required to be treated in the digester.The process is usually run as a batch process with more than one digester tank in operation at any one time. Air is pumped through the tank and the contents are stirred to keep the contents fully mixed. Carbon dioxide, waste air and small quantities of other gases including hydrogen sulfide are given off. These waste gases require treatment to reduce odours in works close to housing or capable of generating public nuisance. The digestion is continued until the percentage is degradable solids is reduced to between 20% and 10% depending on local conditions

ADVANTAGES

Because the aerobic digestion occurs much faster than anaerobic digestion, the capital costs of aerobic digestion are lower.

The process is usually run at ambient temperature and the process is much less complex than anaerobic digestion and is easier to manage.

DISADVANTAGES

The operating costs are typically much greater for aerobic digestion than for anaerobic digestion because of energy used by the blowers, pumps and motors needed to add oxygen to the process. However, recent technological advances include non-electrically aerated filter systems that use natural air currents for the aeration instead of electrically operated machinery. The digested sludge is relatively low in residual energy and although it can be dried and incinerated to produce heat, the energy yield is very much lower than that produced by anaerobic digestion

AEROBIC GRANULATION

The biological treatment of wastewater in the sewage treatment plant is often accomplished using conventional activated sludge systems. These systems generally require large surface areas for treatment and biomass separation units due to the generally poor settling properties of the sludge. In recent years, new technologies have been developed to improve settlability. The use of **aerobic granular sludge technology** is one of them.

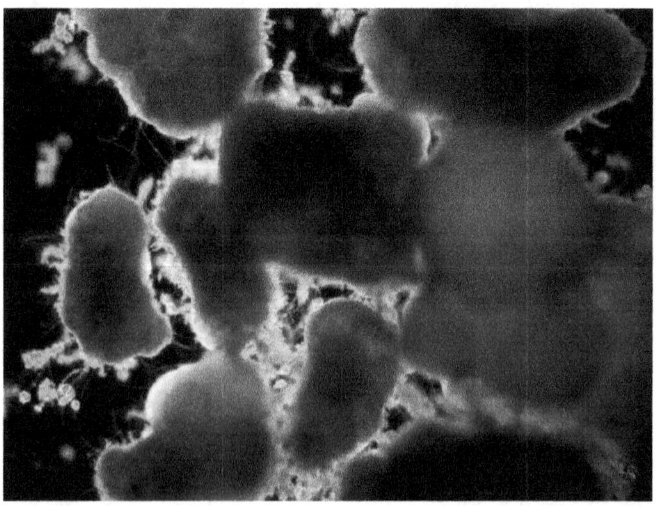

Aerobic Granules derived from municipal sewage AGS application.

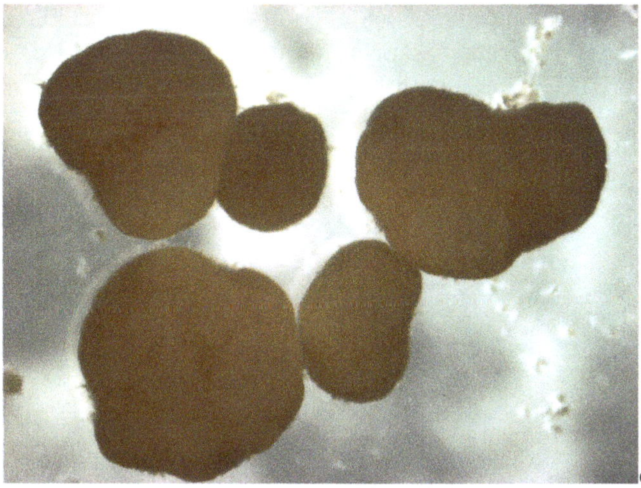
c

Aerobic Granules.

Context

Proponents of aerobic granular sludge technology claim "it will play an important role as an innovative technology alternative to the present activated sludge process in industrial and municipal wastewater treatment in the near future"and that it "can be readily established and profitably used in activated sludge plants". However in 2011 it was characterised as "not yet established as a large-scale application ... with limited and unpublished full-scale applications for municipal wastewater treatment."

Aerobic Granular Biomass

The following definition differentiates an aerobic granule from a simple floc with relatively good settling properties and came out of discussions which took place at the "*1st IWA-Workshop Aerobic Granular Sludge*" in Munich (2004):

"Granules making up aerobic granular activated sludge are to be understood as aggregates of microbial origin, which do not coagulate under reduced hydrodynamic shear, and which settle significantly faster than activated sludge flocs"(de Kreuk *et al.* 2005)"

Formation of Aerobic Granules

Granular sludge biomass is developed in sequencing batch reactors (SBR) and without carrier materials. These systems fulfil most of the requirements for their formation as:

Feast - Famine regime: short feeding periods must be selected to create feast and famine periods (Beun et al. 1999), characterized by the presence or absence of organic matter in the liquid media, respectively. With this feeding strategy the selection of the appropriate micro-organisms to form granules is achieved. When the substrate concentration in the bulk liquid is high, the granule-former organisms can store the organic matter in form of poly-β-hydroxybutyrate to be consumed in the famine period, giving an advantage over filamentous organisms. When an anaerobic feeding is applied this factor is enhanced, minimising the importance of short settling time and higher hydrodynamic forces.

Short settling time: This hydraulic selection pressure on the microbial community allows the retention granular biomass inside the reactor while flocculent biomass is washed-out. (Qin et al. 2004)

SBR Reactor, with aerobic granules.

Granular sludge biomass is developed in sequencing batch reactors (SBR)

Hydrodynamic shear force : Evidences show that the application of high shear forces favours the formation of aerobic granules and the physical granule integrity. It was found that aerobic granules could be formed only above a threshold shear force value in terms of superficial upflow air velocity above 1.2 cm/s in a column SBR, and more regular, rounder, and more compact aerobic granules were developed at high hydrodynamic shear forces (Tay et al., 2001).

Granular activated sludge is also developed in flow-through reactors using the Hybrid Activated Sludge (HYBACS®) process, comprising an attached-growth reactor with short retention time upstream of a suspended growth reactor. The attached bacteria in the first reactor, known as a SMART unit, are exposed to a constant high COD, triggering the expression of high concentrations of hydrolytic enzymes in the EPS layer around the bacteria. The accelerated hydrolysis liberates soluble readily-degradable COD which promotes the formation of granular activated sludge.

Advantages

The development of biomass in the form of aerobic granules is being studied for its application in the removal of organic matter, nitrogen and phosphorus compounds from wastewater. Aerobic granules in an aerobic SBR present several advantages compared to conventional activated sludge process such as:

Stability and flexibility: the SBR system can be adapted to fluctuating conditions with the ability to withstand shock and toxic loadings

Low energy requirements: the aerobic granular sludge process has a higher aeration efficiency due to operation at increased height, while there are neither return sludge or nitrate recycle streams nor mixing and propulsion requirements

Reduced footprint: The increase in biomass concentration that is possible because of the high settling velocity of the aerobic sludge granules and the absence of a final settler result in a significant reduction in the required footprint.

Good biomass retention: higher biomass concentrations inside the reactor can be achieved, and higher substrate loading rates can be treated.

Presence of aerobic and anoxic zones inside the granules: to perform simultaneously different biological processes in the same system (Beun *et al.*. 1999)

Reduced investment and operational costs: the cost of running a wastewater treatment plant working with aerobic granular sludge can be reduced by at least 20% and space requirements can be reduced by as much as 75% (de Kreuk *et al.*., 2004).

The HYBACS process has the additional benefit of being a flow-through process, thus avoiding the complexities of SBR systems. It is also readily applied to the upgrading of existing flow-through activated sludge processes, by installing the attached growth reactors upstream of the aeration tank. Upgrading to granular activated sludge process enables the capacity of an existing wastewater treatment plant to be doubled.

Treatment of Industrial Wastewater

Synthetic wastewater was used in most of the works carried out with aerobic granules. These works were mainly focussed on the study of granules formation, stability and nutrient removal efficiencies under different operational conditions and their potential use to remove toxic compounds. The potential of this technology to treat industrial wastewater is under study, some of the results:

- Arrojo *et al.* (2004) operated two reactors that were fed with industrial wastewater produced in a laboratory for analysis of dairy products (Total COD : 1500–3000 mg/L; soluble COD: 300–1500 mg/L; total nitrogen: 50–200 mg/L). These authors applied organic and nitrogen loading rates up to 7 g COD/(L·d) and 0.7 g N/(L·d) obtaining removal efficiencies of 80%.
- Schwarzenbeck *et al.* (2004) treated malting wastewater which had a high content of particulate organic matter (0.9 g TSS/L). They found that particles with average diameters lower than 25-50 μm were removed at 80% efficiency, whereas particles bigger than 50 μm were only removed at 40% efficiency.

These authors observed that the ability of aerobic granular sludge to remove particulate organic matter from the wastewaters was due to both incorporation into the biofilm matrix and metabolic activity of protozoa population covering the surface of the granules.

- Cassidy and Belia (2005) obtained removal efficiencies for COD and P of 98% and for N and VSS over 97% operating a granular reactor fed with slaughterhouse wastewater (Total COD: 7685 mg/L; soluble COD: 5163 mg/L; TKN: 1057 mg/L and VSS: 1520 mg/L). To obtain these high removal percentages, they operated the reactor at a DO saturation level of 40%, which is the optimal value predicted by Beun *et al.* (2001) for N removal, and with an anaerobic feeding period which helped to maintain the stability of the granules when the DO concentration was limited.
- Inizan *et al.* (2005) treated industrial wastewaters from pharmaceutical industry and observed that the suspended solids in the inlet wastewater were not removed in the reactor.
- Tsuneda *et al.* (2006), when treating wastewater from metal-refinery process (1.0-1.5 g NH_4^+-N/L and up to 22 g/L of sodium sulphate), removed a nitrogen loading rate of 1.0 kg-N/m^3·d with an efficiency of 95% in a system containing autotrophic granules.
- Usmani *et al.* (2008) high superficial air velocity, a relatively short settling time of 5-30 min, a high ratio of height to diameter (H/D=20) of the reactor and optimum organic load facilitates the cultivation of regular compact and circular granules.
- Figueroa *et al.* (2008), treated wastewater from a fish canning industry. Applied OLR were up to 1.72 kg COD/(m^3·d) with fully organic matter depletion. Ammonia nitrogen was removed via nitrification-denitrification up to 40% when nitrogen loading rates were of 0.18 kg N/(m^3·d). The formation of mature aerobic granules occurred after 75 days of operation with 3.4 mm of diameter, SVI of 30 mL/g VSS and density around 60 g VSS/L-granule
- Farooqi *et al.* (2008), Wastewaters from fossil fuel refining, pharmaceuticals, and pesticides are the main sources of phenolic compounds. Those with more complex structures are often more toxic than the simple phenol. This study was aimed at assessing the efficacy of granular sludge in UASB and SBR for the treatment of mixtures of phenolics compounds. The results indicates that anaerobic treatment by UASB and aerobic treatment by SBR can be successfully used for phenol/cresol mixture, representative of major substrates in chemical and petrochemical wastewater and the results shows proper acclimatization period is essential for the degradation of m - cresol and phenol. Moreover, SBR was found as a better alternative than UASB reactor as it is more efficient and higher concentration of m cresols can be successfully degraded.
- López-Palau *et al.* (2009), treated wastewater from a winery industry. The formation of granules was performed using a synthetic substrate and after 120 days of operation, synthetic media was replaced by real winery wastewater, with a COD loading of 6 kg COD/(m^3·d).

Pilot Research in Aerobic Granular Sludge

Aerobic granulation technology for the application in wastewater treatment is widely developed at laboratory scales. The large-scale experience is growing rapidly and multiple institutions are making efforts to improve this technology:

- Since 1999 Royal HaskoningDHV (former DHV Water), Delft University of technology (TUD), STW (Dutch Foundation for Applied Technology) and STOWA (Dutch Foundation for Applied Water Research) have been cooperating closely on the development of the aerobic granular sludge technology (Nereda). In September 2003, a first extensive pilot plant research was executed at STP Ede, the Netherlands with focus on obtaining stable granulation and biological nutrient removal. Following the positive outcome together with six Dutch Water Boards the parties decided to establish a Public-Private Partnership (PPP)- the National Nereda Research Program (NNOP)- to mature, further scale-up and implement several full-scale units. As part of this PPP extensive pilot tests have been executed between 2003 and 2010 at multiple sewage treatment plants. Currently more than 20 plants are running or under construction across 3 continents.
- From the basis of the aerobic granular sludge but using a contention system for the granules, a sequencing batch biofilter granular reactor (SBBGR)3 with a volume of $3.1m^3$ was developed by IRSA (Istituto di Ricerca Sulle Acque, Italy). Different studies were carried out in this plant treating sewage at an Italian wastewater treatment plant.
- The use of aerobic granules prepared in laboratory, as a starter culture, before adding in main system, is the base of the technology ARGUS (Aerobic granules upgrade system) developed by EcoEngineering Ltd.. The granules are cultivated on-site in small bioreactors called propagators and fill up only 2 to 3% of the main bioreactor or fermentor (digestor) capacity. This system is being used in a pilot plant with a volume of $2.7\ m^3$ located in one Hungarian pharmaceutical industry.
- The Group of Environmental Engineering and Bioprocesses from the University of Santiago de Compostela is currently operating a 100 L pilot plant reactor.

The feasibility study showed that the aerobic granular sludge technology seems very promising (de Bruin et al., 2004. Based on total annual costs a GSBR (Granular sludge sequencing batch reactors) with pre-treatment and a GSBR with post-treatment proves to be more attractive than the reference activated sludge alternatives (6-16%). A sensitivity analysis shows that the GSBR technology is less sensitive to land price and more sensitive to rain water flow. Because of the high allowable volumetric load the footprint of the GSBR variants is only 25% compared to the references. However, the GSBR with only primary treatment cannot meet the present effluent standards for municipal wastewater, mainly because of exceeding the suspended solids effluent standard caused by washout of not well settleable biomass.

Full Scale Application

Aerobic granulation technology is already successfully applied for treatment of wastewater.

- Since 2005, RoyalHaskoningDHV has implemented more than 20 full-scale aerobic granular sludge technology systems (Nereda) for the treatment of both industrial and municipal wastewater across 3 continents. One example is STP Epe, The Netherlands, with a capacity of 59.000 pe and 1,500 m3.h-1, being the first full-scale municipal Nereda in The Netherlands. Examples of the latest Nereda sewage treatment plants (2012-2013) include Wemmershoek- South Africa, Dinxperlo, Vroomshoop, Garmerwolde - The Netherlands.

Full-scale municipal sewage Nereda application (4000 m3.d-1) at the Gansbaai STP in South Africa.

Full-scale municipal sewage Nereda application Epe the Netherlands.

Full-scale industrial sewage Nereda application Vika the Netherlands.

- EcoEngineering applied aerobic granulation process in three pharmaceutical industries, Krka d.d. Novo mesto Slovenia, Lek d.d. Lendava, Slovenia and Gedeon Richter Rt. Dorog, Hungary. Wastewater treatment plants are already running more than five years.

AEROBIC GRANULAR REACTOR

Aerobic granular reactor is a type of waste treatment facility.

Sewage treatment plants (STP) based on activated sludge often cover large surface areas, necessitated mainly by the large settling tanks. To build compact STP's, biomass can be grown as biofilms on a carrier material, or as fast settling aerobic granular sludge without a carrier. Recent research showed the advantages of a discontinuously fed system, in which it is possible to grow stable granulated sludge under aerobic conditions.

Simultaneous Chemical oxygen demand, and Nitrogen and Phosphorus removal, can be easily integrated in a discontinuous fed system. Because of the high settling capacity of the granules, the use of a traditional settler is unnecessary. Therefore the installation can be built very compact, needing only 20% of the surface area of conventional activated sludge systems.

BIOREACTOR

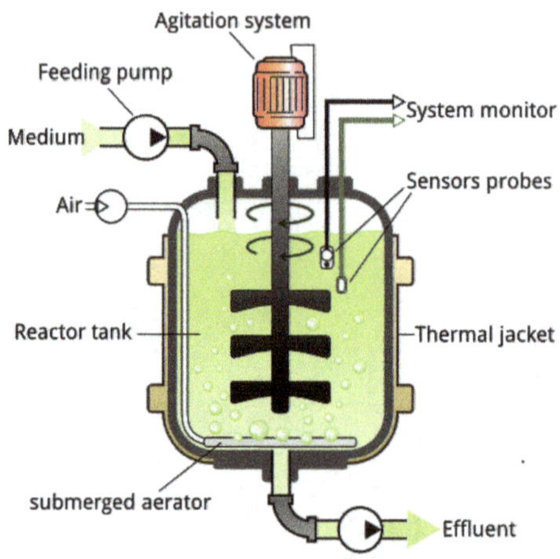

General structure of a continuous stirred-tank type bioreactor.

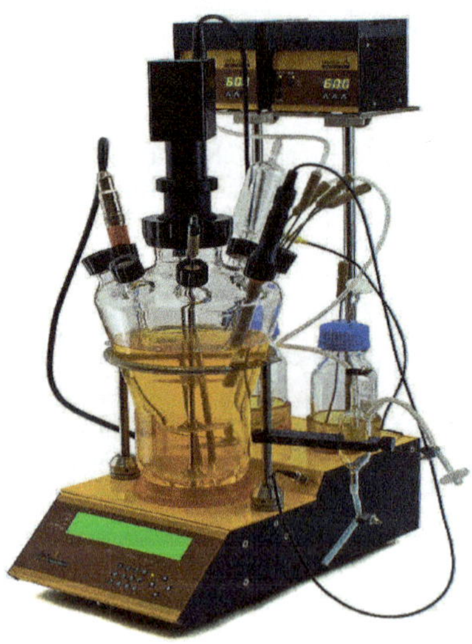

Autoclavable bench-top laboratory bioreactor used for fermentation and cell cultures.

A **bioreactor** may refer to any manufactured or engineered device or system that supports a biologically active environment. In one case, a bioreactor is a vessel in which a chemical process is carried out which involves organisms or biochemically active substances derived from such organisms. This process can either be aerobic or anaerobic. These bioreactors are commonly cylindrical, ranging in size from litres to cubic metres, and are often made of stainless steel.

A bioreactor may also refer to a device or system meant to grow cells or tissues in the context of cell culture. These devices are being developed for use in tissue engineering or biochemical engineering.

On the basis of **mode of operation**, a bioreactor may be classified as batch, fed batch or continuous (e.g. a continuous stirred-tank reactor model). An example of a continuous bioreactor is the chemostat.

Organisms growing in bioreactors may be submerged in liquid medium or may be attached to the surface of a solid medium. Submerged cultures may be suspended or immobilized. Suspension bioreactors can use a wider variety of organisms, since special attachment surfaces are not needed, and can operate at much larger scale than immobilized cultures. However, in a continuously operated process the organisms will be removed from the reactor with the effluent. Immobilization is a general term describing a wide variety of cell or particle attachment or entrapment. It can be applied to basically all types of biocatalysis including enzymes, cellular organelles, animal and plant cells. Immobilization is useful for continuously operated processes, since the organisms will not be removed with the reactor effluent, but is limited in scale because the microbes are only present on the surfaces of the vessel.

Large scale immobilized cell bioreactors are:

- moving media, also known as moving bed biofilm reactor (MBBR)
- packed bed
- fibrous bed
- membrane

Bioreactor Design

Bioreactor design is a relatively complex engineering task, which is studied in the discipline of biochemical engineering. Under optimum conditions, the microorganisms or cells are able to perform their desired function with limited production of impurities. The environmental conditions inside the bioreactor, such as temperature, nutrient concentrations, pH, and dissolved gases (especially oxygen for aerobic fermentations) affect the growth and productivity of the organisms. The temperature of the fermentation medium is maintained by a cooling jacket, coils, or both. Particularly exothermic fermentations may require the use of external heat exchangers. Nutrients may be continuously added to the fermenter, as in a fed-batch system, or may be charged into the reactor at the beginning of fermentation. The pH of the medium is measured and adjusted with small amounts of acid or base, depending upon the fermentation. For aerobic (and some anaerobic) fermentations, reactant gases (especially oxygen) must be added to the fermentation. Since oxygen is relatively insoluble in water (the basis of nearly all fermentation media), air (or purified oxygen) must be added continuously. The action of the rising bubbles helps mix the fermentation medium and also “strips” out waste gases, such as carbon dioxide.

In practice, bioreactors are often pressurized; this increases the solubility of oxygen in water. In an aerobic process, optimal oxygen transfer is sometimes the rate limiting step. Oxygen is poorly soluble in water—even less in warm fermentation broths—and is relatively scarce in air (20.95%). Oxygen transfer is usually helped by agitation, which is also needed to mix nutrients and to keep the fermentation homogeneous. Gas dispersing agitators are used to break up air bubbles and circulate them throughout the vessel.

A closed bioreactor used in cellulosic ethanol research.

Fouling can harm the overall efficiency of the bioreactor, especially the heat exchangers. To avoid it, the bioreactor must be easily cleaned. Interior surfaces are typically made of stainless steel for easy cleaning and sanitation. Typically bioreactors are cleaned between batches, or are designed to reduce fouling as much as possible when operated continuously. Heat transfer is an important part of bioreactor design; small vessels can be cooled with a cooling jacket, but larger vessels may require coils or an external heat exchanger.

Types of Bioreactors

Photobioreactor

Moss photobioreactor with *Physcomitrella patens*.

A photobioreactor (PBR) is a bioreactor which incorporates some type of light source (that may be natural sunlight or artificial illumination). Virtually any translucent container could be called a PBR, however the term is more commonly used to define a closed system, as opposed to an open tank or pond. Photobioreactors are used to grow small phototrophic organisms such as cyanobacteria, algae, or moss plants. These organisms use light through photosynthesis as their energy source and do not require sugars or lipids as energy source. Consequently, risk of contamination with other organisms like bacteria or fungi is lower in photobioreactors when compared to bioreactors for heterotroph organisms.

Sewage Treatment

Bioreactors are also designed to treat sewage and wastewater. In the most efficient of these systems, there is a supply of a free-flowing, chemically inert medium which acts as a receptacle for the bacteria that break down the raw sewage. Examples of these bioreactors often have separate, sequential tanks and a mechanical separator or cyclone to speed the separation of water and biosolids. Aerators supply oxygen to the sewage and medium, further accelerating breakdown. Submersible mixers provide agitation in anoxic bioreactors to keep the solids in suspension and thereby ensure that the bacteria and the organic materials "meet". In the process, the liquid's Biochemical Oxygen Demand (BOD) is reduced sufficiently to render the contaminated water fit for reuse. The biosolids can be collected for further processing, or dried and used as fertilizer. An extremely simple version of a sewage bioreactor is a septic tank whereby the sewage is left in situ, with or without additional media to house bacteria. In this instance, the biosludge itself is the primary host (activated sludge) for the bacteria. Septic systems are best suited where there is sufficient landmass, and the system is not subject to flooding or overly saturated ground, and where time and efficiency are not prioritized.

Because they are the engine that drives biological wastewater treatment, it is critical to closely monitor the quantity and quality of microorganisms in bioreactors. One method for this is via 2nd Generation ATP tests.

Up and down Agitation Bioreactor

Up and down agitators are useful to avoid shear stress to the cells. These are done by instead of a traditional propeller agitator, which requires an expensive motor and magnetic coupling. Vertical up and down motion is achieved by a motor together with an inexpensive membrane perfectly assure sterility and produce an efficient mixing without formation of a vortex (no baffles needed). At the same time this type of mixing is gentler on cells and produces less foam. Novel biomimicking "fish-tail" stirring discs offer maximum mixing efficiency without cutting edges.

NASA Tissue Cloning Bioreactor

A bioreactor used to ferment ethanol from corncob waste being loaded with yeast.

In bioreactors in which the goal is to grow cells or tissues for experimental or therapeutic purposes, the design is significantly different from industrial bioreactors. Many cells and tissues, especially mammalian ones, must have a surface or other structural support in order to grow, and agitated environments are often destructive to these cell types and tissues. Higher organisms, being auxotrophic, also require highly specialized growth media.

NASA has developed a new type of bioreactor that artificially grows tissue in cell cultures. NASA's tissue bioreactor can grow heart tissue, skeletal tissue, ligaments, cancer tissue for study, and other types of tissue.

For more information on artificial tissue culture, see tissue engineering.

Modelling of bioreactor

Mathematical models act as an important tool in various bio-reactor applications including wastewater treatment. These models are useful for planning efficient process control strategies and predicting the future plant performance. Moreover, these models are beneficial in education and research areas.

Bioreactors are generally used in those industries which are concerned with food, beverages and pharmaceuticals. The emergence of *Biochemical engineering* is of recent origin. Processing of biological materials using biological agents such as cells, enzymes or antibodies are the major pillars of biochemical engineering. Applications of biochemical engineering cover major fields of civilization such as agriculture, food and healthcare, resource recovery and fine chemicals.

Till now, the industries associated with biotechnology have been lagged behind other industries in implementing control over the process and optimization strategies. A main drawback in biotechnological process control is the problem to measure key physical and biochemical parameters.

Operational Stages in a Bio-process

A bioprocess is composed mainly of three stages — upstream processing, bioreaction, and downstream processing — to convert raw material to finished product.

The raw material can be of biological or non-biological origin. It is first converted to more suitable form for processing. This is done in upstream processing step which involves chemical hydrolysis, preparation of liquid medium, separation of particulate, air purification and many other preparatory operations.

After upstream processing step, the resulting feed is transferred to one or more Bioreaction stages. The Biochemical reactors or bioreactors form the base of the Bioreaction step. This step is mainly consists of three operations namely, production of biomass, metabolize biosynthesis and biotransformation.

Finally, the material produced in the bioreactor must be further processed in the downstream section to convert it into more useful form. The downstream process is mainly consists of physical separation operations which includes, solid liquid separation, adsorption, liquid-liquid extraction, distillation, drying etc.

Specifications of a Bioreactor

A typical bioreactor consists of following parts:

Agitator – used for the mixing of the contents of the reactor which keeps the "cells" in the perfect homogenous condition for better transport of nutrients and oxygen to the desired product(s).

Baffle – used to break the vortex formation in the vessel, which is usually highly undesirable as it changes the center of gravity of the system and consumes additional power.

Sparger – In aerobic cultivation process, the purpose of the sparger is to supply adequate oxygen to the growing cells.

Jacket – The jacket provides the annular area for circulation of constant temperature of water which keeps the temperature of the bioreactor at a constant value.

Development of Modelling Equations for Bioreactors

Assumptions –

- The reactor contents are perfectly mixed together.
- The reactor is operating at a constant temperature (i.e., it is isothermal).
- The feed should be clean and pure (i.e., no biomass in the feed stream).
- The feed stream and reactor contents have equal and constant density (ρ).
- The feed and product streams have the same flow rate (F).

Total Continuity Equation

Making overall mass balance, we get the following equation:

$d(\rho V)/dt = F\rho - F\rho = 0$ (1)

Equation(1) states that the reactor volume (V) is constant since $dV/dt = 0$.

Biomass Continuity Equation

We know,

Flow rate of biomass into the reactor = Fx_i

Flow rate of biomass out of the reactor = Fx

Rate of generation of biomass by reaction = Vr_1

Rate of accumulation of biomass within the reactor = $d(Vx)/dt$

Now, apply general mass balance equation i.e,

Rate of Mass In – Rate of Mass Out + Rate of Generation = Accumulation

$d(Vx)/dt = Fx_i - Fx + Vr_1$ (2)

Where $\mathbf{r_1}$ is the rate of cell generation. Dividing both sides of the above equation by V, we obtain

$dx/dt = (F/V)x_i - (F/V)x + r_1$ (3)

In the chemical reaction engineering, **F/V** is called **space velocity**(s^{-1}) and **V/F** is called the **residence time** (s). But in biochemical engineering, **F/V** is known as **Dilution rate (D_r)**. Accordingly, equation(3) yields:

$dx/dt = D_r x_i - D_r x + r_1$ (4)

$dx/dt = D_r(x_i - x) + r_1$ (5)

Substrate Continuity Equation

For substrate balance,

Flow rate of substrate into the bioreactor = FS_i

Flow of the substrate out of the bioreactor = FS

Rate of generation of substrate by reaction = $-Vr_2$

Rate of accumulation of substrate within the reactor = d(VS)/dt

Now, apply general mass balance equation i.e,

Rate of Mass In – Rate of Mass Out + Rate of Generation = Accumulation

$d(VS)/dt = FS_i - FS - Vr_2$ (6)

rearranging above equation, we get

$dS/dt = D_r(S_i - S) - r_2$ (7)

where $\mathbf{r_2}$ is the rate of substrate consumption.

Biochemical Reaction Kinetics

For the chemical reaction,

A ----> P

We can write

$(-r_A) = k\,(C_A)^n$ (8)

$(r_A) = -k\,(C_A)^n$ (9)

Where,

$(-r_A)$ = rate of disappearance of A

(r_A) = rate of formation of A

k = reaction rate constant

C_A = Concentration of reactant A

n = order of reaction with respect to component A

For first order reaction, n = 1 and accordingly,

$-r_A = k\,C_A$

The reaction kinetics involved in biochemical operations is comparatively difficult to obtain than the chemical reaction kinetics. In biochemical operations, the cell kinetics is used for the unstructured models where balanced growth condition is assumed.

The following equation is used to represent the net rate of cell mass growth:

$r_1 = \mu x$ *(10)*

where μ is the specific growth rate or specific growth rate coefficient(s^{-1}). Here, μ is analogous to first order rate constant k but however, μ is not a constant.

In biochemical engineering, **yield** is defined as the ratio of mass or moles of product formed to the mass or moles of the reactants consumed. The yield (Y) of product (P) with respect to reactant A is defined as:

Y = (mass of P formed)/(mass of A consumed) (11)

In case of bioreactor,

Y = (mass of cells formed)/(mass of substrate consumed) (12)

Thus,

$Y = r_1/ r_2$

Or,

$r_2 = r_1/Y$

Or,

$r_2 = \mu x/Y$ (from 10) (13)

Final Form of Equation of Modelling

By substituting equations (10) & (13) in equations (5) & (7) respectively, we get,

$\mathbf{dx/dt = D_r(x_i - x) + \mu x}$ (14)

$\mathbf{dS/dt = S(S_i - S) - (\mu x/Y)}$ (15)

Since we have assumed that the feed stream does not contain any biomass i.e., $\mathbf{x_i = 0}$, then, bioreactor modelling equation finally get the following form:

$\mathbf{dx/dt = (\mu - D_r)x}$ (16)

$\mathbf{dS/dt = S(S_i - S) - (\mu x/Y)}$ *(from 15)*

Thus,Equations (15) and (16) are the basic equations which are used for the modelling of any bioreactor.

Distillation is a process of separating the component or substances from a liquid mixture by selective evaporation and condensation. Distillation may result in essentially complete separation (nearly purified components), or it may be a partial separation that increases the concentration of selected components of the mixture. In either case the process exploits differences in the volatility of the mixture's components. In industrial chemistry, distillation is a unit operation of practically universal importance, but it is a physical separation process and not a chemical reaction.

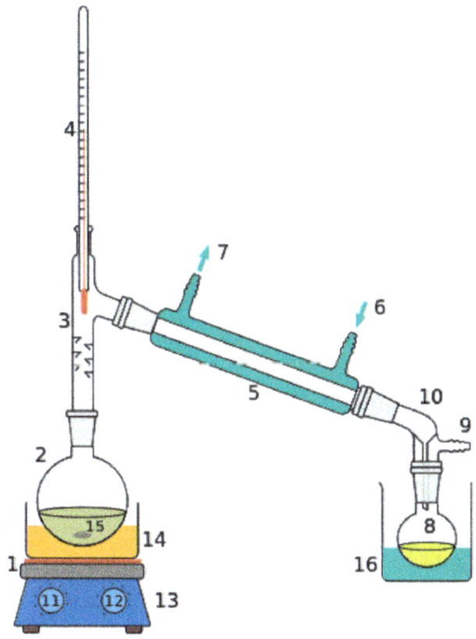

Laboratory display of distillation: **1:** A source of heat **2:** Still pot **3:** Still head **4:** Thermometer/ Boiling point temperature **5:** Condenser **6:** Cooling water in **7:** Cooling water out **8:** Distillate/ receiving flask **9:** Vacuum/gas inlet **10:** Still receiver **11:** Heat control **12:** Stirrer speed control **13:** Stirrer/heat plate **14:** Heating (Oil/sand) bath **15:** Stirring means e.g. (shown), boiling chips or mechanical stirrer **16:** Cooling bath.

Commercially, distillation has many applications. For example:

- In the fossil fuel industry distillation is a major class of operation in obtaining materials from crude oil for fuels and for chemical feed stocks.
- Distillation permits separation of air into its components — notably oxygen, nitrogen, and argon — for industrial use.
- In the field of industrial chemistry, large ranges of crude liquid products of chemical synthesis are distilled to separate them, either from other products, or from impurities, or from unreacted starting materials.
- Distillation of fermented products produces distilled beverages with a high alcohol content, or separates out other fermentation products of commercial value.

An installation for distillation, especially of alcohol, is a **distillery**. The distillation equipment is a **still**.

Distillation is a very old method of artificial desalination.

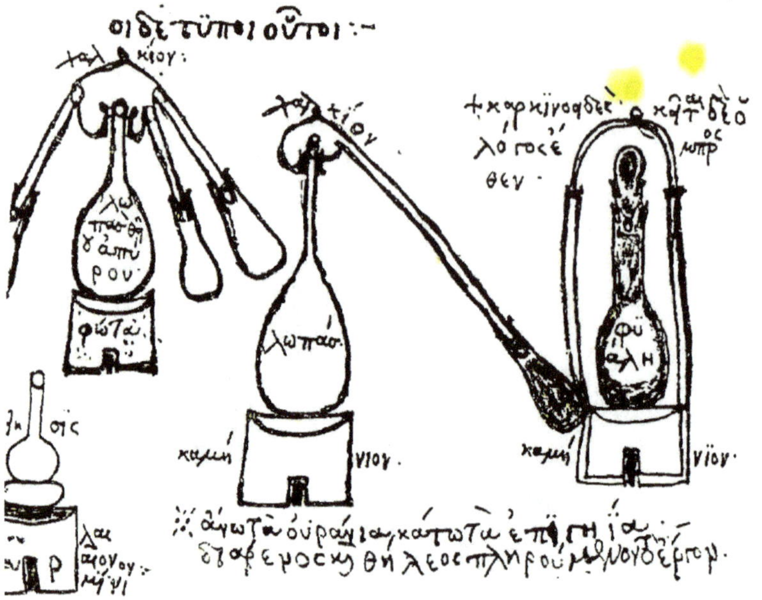

Distillation equipment used by the 3rd century Greek alchemist Zosimos of Panopolis, from the Byzantine Greek manuscript *Parisinus graces*.

Aristotle wrote about the process in his *Meteorologica* and even that "ordinary wine possesses a kind of exhalation, and that is why it gives out a flame". Later evidence of distillation comes from Greek alchemists working in Alexandria in the 1st century AD. Distilled water has been known since at least c. 200, when Alexander of Aphrodisias described the process.Work on distilling other liquids continued in early Byzantine Egypt under the Greek-Egyptian Zosimus of Panopolis. Distillation in China could have begun during the Eastern Han Dynasty (1st–2nd centuries), but archaeological evidence indicates that actual distillation of beverages began in the Jin (12th–13th centuries) and Southern Song (10th–13th centuries) dynasties. A still was found in an archaeological site in Qinglong, Hebei province dating to the 12th century. Distilled beverages were more common during the Yuan dynasty (13th–14th centuries). Arabs learned the process from the Alexandrians and used it extensively in their chemical experiments.

Clear evidence of the distillation of alcohol comes from the School of Salerno in the 12th century. Fractional distillation was developed by Tadeo Alderotti in the 13th century.

In 1500, German alchemist Hieronymus Braunschweig published *Liber de arte destillandi* (The Book of the Art of Distillation) the first book solely dedicated to the subject of distillation, followed in 1512 by a much expanded version. In 1651, John French published The Art of Distillation the first major English compendium of practice, though it has been claimed that much of it derives from Braunschweig's work. This includes diagrams with people in them showing the industrial rather than bench scale of the operation.

Hieronymus Brunschwig's *Liber de arte Distillandi de Compositis* (Strassburg, 1512) Chemical Heritage Foundation.

A retort

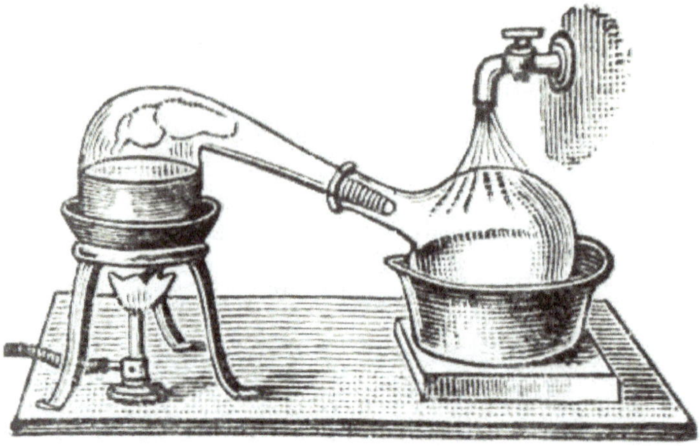

Distillation

Old Ukrainian vodka still

Simple liqueur distillation in East Timor.

As alchemy evolved into the science of chemistry, vessels called retorts became used for distillations. Both alembics and retorts are forms of glassware with long necks pointing to the side at a downward angle which acted as air-cooled condensers to condense the distillate and let it drip downward for collection. Later, copper alembics were invented. Riveted joints were often kept tight by using various mixtures, for instance a dough made of rye flour. These alembics often featured a cooling system around the beak, using cold water for instance, which made the condensation of alcohol more efficient. These were called pot stills. Today, the retorts and pot stills have been largely supplanted by more efficient distillation methods in most industrial processes. However, the pot still is still widely used for the elaboration of some fine alcohols such as cognac, Scotch whisky, tequila and some vodkas. Pot stills made of various materials (wood, clay, stainless steel) are also used by bootleggers in various countries. Small pot stills are also sold for the domestic production of flower water or essential oils.

Early forms of distillation were batch processes using one vaporization and one condensation. Purity was improved by further distillation of the condensate. Greater volumes were processed by simply repeating the distillation. Chemists were reported to carry out as many as 500 to 600 distillations in order to obtain a pure compound.

In the early 19th century the basics of modern techniques including pre-heating and reflux were developed. In 1822, Anthony Perrier developed one of the first continuous stills. In 1826, Robert Stein improved that design to make his patent still. In 1830, Aeneas Coffey got a patent for improving that design. Coffey's continuous still may be regarded as the archetype of modern petrochemical units. The French engineer Armand Savalle developed his steam regulator around 1846. In 1877, Ernest Solvay was granted a U.S. Patent for a tray column for ammonia distillation and the same and subsequent years saw developments of this theme for oil and spirits.

With the emergence of chemical engineering as a discipline at the end of the 19th century, scientific rather than empirical methods could be applied. The developing petroleum industry in the early 20th century provided the impetus for the development of accurate design methods such as the McCabe–Thiele method and the Fenske equation. The availability of powerful computers has also allowed direct computer simulation of distillation columns.

Applications of Distillation

The application of distillation can roughly be divided in four groups: laboratory scale, industrial distillation, distillation of herbs for perfumery and medicinals (herbal distillate), and food processing. The latter two are distinctively different from the former two in that in the processing of beverages and herbs, the distillation is not used as a true purification method but more to transfer all volatiles from the source materials to the distillate.

The main difference between laboratory scale distillation and industrial distillation is that laboratory scale distillation is often performed batch-wise, whereas industrial distillation often occurs continuously. In batch distillation, the composition of the source material, the vapors of the distilling compounds and the distillate change during the distillation. In batch distillation, a still is charged (supplied) with a batch of feed mixture, which is then separated into its component fractions which are collected sequentially from most volatile to less volatile, with the bottoms (remaining least or non-volatile fraction) removed at the end. The still can then be recharged and the process repeated.

In continuous distillation, the source materials, vapors, and distillate are kept at a constant composition by carefully replenishing the source material and removing fractions from both vapor and liquid in the system. This results in a better control of the separation process.

Idealized Distillation Model

The boiling point of a liquid is the temperature at which the vapor pressure of the liquid equals the pressure around the liquid, enabling bubbles to form without being crushed. A special case is the normal boiling point, where the vapor pressure of the liquid equals the ambient atmospheric pressure.

It is a common misconception that in a liquid mixture at a given pressure, each component boils at the boiling point corresponding to the given pressure and the vapors of each component will collect separately and purely. This, however, does not occur even in an idealized system. Idealized models of distillation are essentially governed by Raoult's law and Dalton's law, and assume that vapor–liquid equilibria are attained.

Raoult's law states that the vapor pressure of a solution is dependent on 1) the vapor pressure of each chemical component in the solution and 2) the fraction of solution each component makes up a.k.a. the mole fraction. This law applies to ideal solutions, or solutions that have different components but whose molecular interactions are the same as or very similar to pure solutions.

Dalton's law states that the total pressure is the sum of the partial pressures of each individual component in the mixture. When a multi-component liquid is heated, the vapor pressure of each component will rise, thus causing the total vapor pressure to rise. When the total vapor pressure reaches the pressure surrounding the liquid, boiling occurs and liquid turns to gas throughout the bulk of the liquid. Note that a mixture with a given composition has one boiling point at a given pressure, when the components are mutually soluble. A mixture of constant composition does not have multiple boiling points.

An implication of one boiling point is that lighter components never cleanly boil first". At boiling point, all volatile components boil, but for a component, its percentage in the vapor is the same as its percentage of the total vapor pressure. Lighter components have a higher partial pressure and thus are concentrated in the vapor, but heavier volatile components also have a (smaller) partial pressure and necessarily evaporate also, albeit being less concentrated in the vapor. Indeed, batch distillation and fractionation succeed by varying the composition of the mixture. In batch distillation, the batch evaporates, which changes its composition; in fractionation, liquid higher in the fractionation column contains more lights and boils at lower temperatures. Therefore, starting from a given mixture, it appears to have a **boiling range** instead of a boiling *point*, although this is because its composition changes: each intermediate mixture has its own, singular boiling point.

The idealized model is accurate in the case of chemically similar liquids, such as benzene and toluene. In other cases, severe deviations from Raoult's law and Dalton's law are observed, most famously in the mixture of ethanol and water. These compounds, when heated together, form an azeotrope, which is a composition with a boiling point higher or lower than the boiling point of each separate liquid. Virtually all liquids, when mixed and heated, will display azeotropic behaviour. Although there are computational methods that can be used to estimate the behavior of a mixture of arbitrary components, the only way to obtain accurate vapor–liquid equilibrium data is by measurement.

It is not possible to *completely* purify a mixture of components by distillation, as this would require each component in the mixture to have a zero partial pressure. If ultra-pure products are the goal, then further chemical separation must be applied. When a binary mixture is evaporated and the other component, e.g. a salt, has zero partial pressure for practical purposes, the process is simpler and is called evaporation in engineering.

Batch Distillation

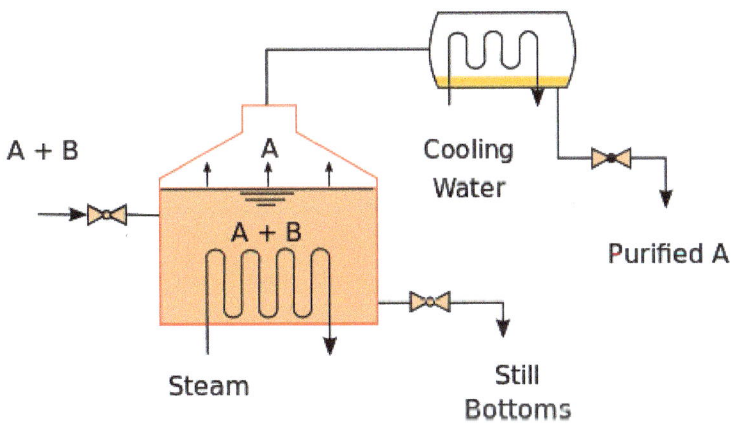

A batch still showing the separation of A and B.

Heating an ideal mixture of two volatile substances A and B (with A having the higher volatility, or lower boiling point) in a batch distillation setup (such as in an apparatus depicted in the opening figure) until the mixture is boiling results in a vapor above the liquid which contains a mixture of A and B. The ratio between A and B in the vapor will be different from the ratio in the liquid: the ratio in the liquid will be determined by how the original mixture was prepared, while the ratio in the vapor will be enriched in the more volatile compound, A (due to Raoult's Law, see above). The vapor goes through the condenser and is removed from the system. This in turn means that the ratio of compounds in the remaining liquid is now different from the initial ratio (i.e., more enriched in B than the starting liquid).

The result is that the ratio in the liquid mixture is changing, becoming richer in component B. This causes the boiling point of the mixture to rise, which in turn results in a rise in the temperature in the vapor, which results in a changing ratio of A : B in the gas phase (as distillation continues, there is an increasing proportion of B in the gas phase). This results in a slowly changing ratio A : B in the distillate.

If the difference in vapor pressure between the two components A and B is large (generally expressed as the difference in boiling points), the mixture in the beginning of the distillation is highly enriched in component A, and when component A has distilled off, the boiling liquid is enriched in component B.

Continuous Distillation

Continuous distillation is an ongoing distillation in which a liquid mixture is continuously (without interruption) fed into the process and separated fractions are removed continuously as output streams occur over time during the operation. Continuous distillation produces a minimum of two output fractions, including at least one volatile distillate fraction, which has boiled and been separately captured as a vapor, and then condensed to a liquid. There is always a bottoms (or residue) fraction, which is the least volatile residue that has not been separately captured as a condensed vapor.

Continuous distillation differs from batch distillation in the respect that concentrations should not change over time. Continuous distillation can be run at a steady state for an arbitrary amount of time. For any source material of specific composition, the main variables that affect the purity of products in continuous distillation are the reflux ratio and the number of theoretical equilibrium stages, in practice determined by the number of trays or the height of packing. Reflux is a flow from the condenser back to the column, which generates a recycle that allows a better separation with a given number of trays. Equilibrium stages are ideal steps where compositions achieve vapor–liquid equilibrium, repeating the separation process and allowing better separation given a reflux ratio. A column with a high reflux ratio may have fewer stages, but it refluxes a large amount of liquid, giving a wide column with a large holdup. Conversely, a column with a low reflux ratio must have a large number of stages, thus requiring a taller column.

General Improvements

Both batch and continuous distillations can be improved by making use of a fractionating column on top of the distillation flask. The column improves separation by providing a larger surface area for the vapor and condensate to come into contact. This helps it remain at equilibrium for as long as possible. The column can even consist of small subsystems ('trays' or 'dishes') which all contain an enriched, boiling liquid mixture, all with their own vapor–liquid equilibrium.

There are differences between laboratory-scale and industrial-scale fractionating columns, but the principles are the same. Examples of laboratory-scale fractionating columns (in increasing efficiency) include

- Air condenser
- Vigreux column (usually laboratory scale only)
- Packed column (packed with glass beads, metal pieces, or other chemically inert material)
- Spinning band distillation system.

Laboratory Scale Distillation

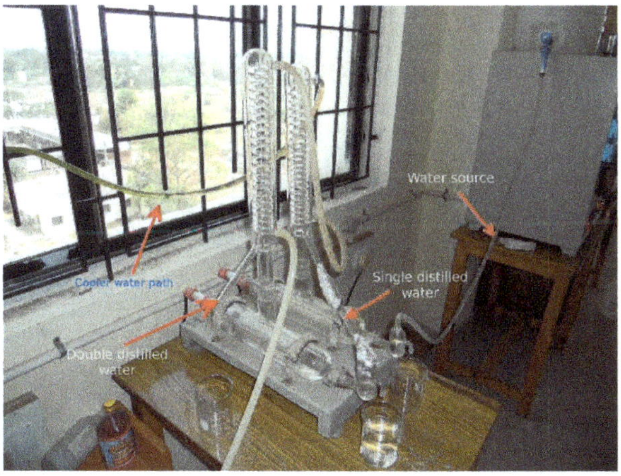

Typical laboratory distillation unit.

Laboratory scale distillations are almost exclusively run as batch distillations. The device used in distillation, sometimes referred to as a *still*, consists at a minimum of a **reboiler** or *pot* in which the source material is heated, a **condenser** in which the heated vapour is cooled back to the liquid state, and a **receiver** in which the concentrated or purified liquid, called the **distillate**, is collected. Several laboratory scale techniques for distillation exist (see also distillation types).

Simple Distillation

In **simple distillation**, the vapor is immediately channeled into a condenser. Consequently, the distillate is not pure but rather its composition is identical to the composition of the vapors at the given temperature and pressure. That concentration follows Raoult's law.

As a result, simple distillation is effective only when the liquid boiling points differ greatly (rule of thumb is 25 °C) or when separating liquids from non-volatile solids or oils. For these cases, the vapor pressures of the components are usually different enough that the distillate may be sufficiently pure for its intended purpose.

Fractional Distillation

For many cases, the boiling points of the components in the mixture will be sufficiently close that Raoult's law must be taken into consideration. Therefore, **fractional distillation** must be used in order to separate the components by repeated vaporization-condensation cycles within a packed fractionating column. This separation, by successive distillations, is also referred to as **rectification**.

As the solution to be purified is heated, its vapors rise to the fractionating column. As it rises, it cools, condensing on the condenser walls and the surfaces of the packing material. Here, the condensate continues to be heated by the rising hot vapors; it vaporizes once more. However, the composition of the fresh vapors are determined once again by Raoult's law. Each vaporization-condensation cycle (called a *theoretical plate*) will yield a purer solution of the more volatile component. In reality, each cycle at a given temperature does not occur at exactly the same position in the fractionating column; *theoretical plate* is thus a concept rather than an accurate description.

More theoretical plates lead to better separations. A spinning band distillation system uses a spinning band of Teflon or metal to force the rising vapors into close contact with the descending condensate, increasing the number of theoretical plates.

Steam Distillation

Like vacuum distillation, **steam distillation** is a method for distilling compounds which are heat-sensitive. The temperature of the steam is easier to control than the surface of a heating element, and allows a high rate of heat transfer without heating at a very high temperature. This process involves bubbling steam through a heated mixture of the raw material. By Raoult's law, some of the target compound will vaporize (in accordance with its partial pressure). The vapor mixture is cooled and condensed, usually yielding a layer of oil and a layer of water.

Steam distillation of various aromatic herbs and flowers can result in two products; an essential oil as well as a watery herbal distillate. The essential oils are often used in perfumery and aromatherapy while the watery distillates have many applications in aromatherapy, food processing and skin care.

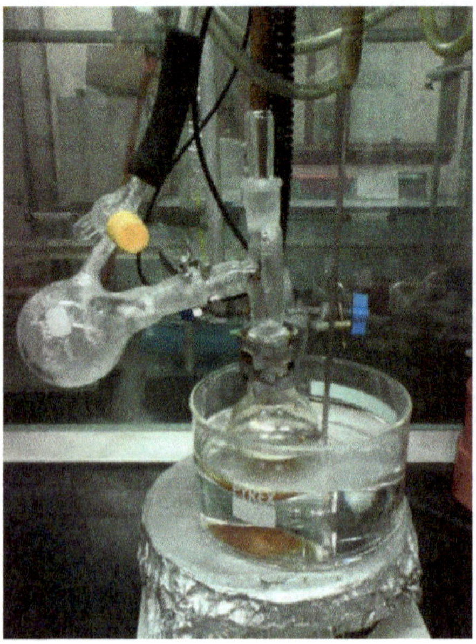

Dimethyl sulfoxide usually boils at 189 °C. Under a vacuum, it distills off into the receiver at only 70 °C.

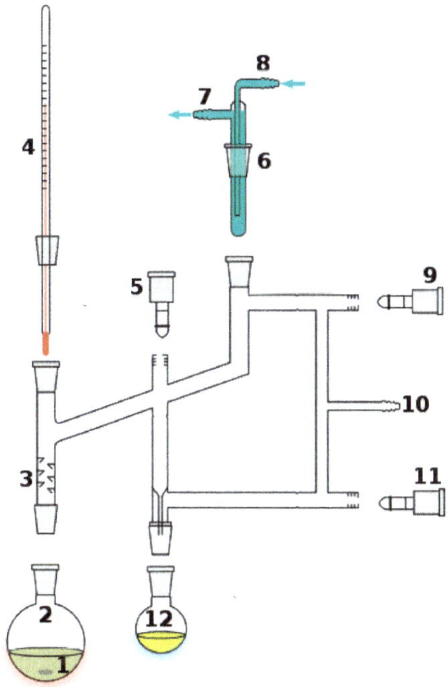

Perkin triangle distillation setup

1: Stirrer bar/anti-bumping granules **2:** Still pot **3:** Fractionating column **4:** Thermometer/Boiling point temperature **5:** Teflon tap 1 **6:** Cold finger **7:** Cooling water out **8:** Cooling water in **9:** Teflon tap 2 **10:** Vacuum/gas inlet **11:** Teflon tap 3 **12:** Still receiver

Vacuum Distillation

Some compounds have very high boiling points. To boil such compounds, it is often better to lower the pressure at which such compounds are boiled instead of increasing the temperature. Once the pressure is lowered to the vapor pressure of the compound (at the given temperature), boiling and the rest of the distillation process can commence. This technique is referred to as **vacuum distillation** and it is commonly found in the laboratory in the form of the rotary evaporator.

This technique is also very useful for compounds which boil beyond their decomposition temperature at atmospheric pressure and which would therefore be decomposed by any attempt to boil them under atmospheric pressure.

Molecular distillation is vacuum distillation below the pressure of 0.01 torr. 0.01 torr is one order of magnitude above high vacuum, where fluids are in the free molecular flow regime, i.e. the mean free path of molecules is comparable to the size of the equipment. The gaseous phase no longer exerts significant pressure on the substance to be evaporated, and consequently, rate of evaporation no longer depends on pressure. That is, because the continuum assumptions of fluid dynamics no longer apply, mass transport is governed by molecular dynamics rather than fluid dynamics. Thus, a short path between the hot surface and the cold surface is necessary, typically by suspending a hot plate covered with a film of feed next to a cold plate with a line of sight in between. Molecular distillation is used industrially for purification of oils.

Air-sensitive Vacuum Distillation

Some compounds have high boiling points as well as being air sensitive. A simple vacuum distillation system as exemplified above can be used, whereby the vacuum is replaced with an inert gas after the distillation is complete. However, this is a less satisfactory system if one desires to collect fractions under a reduced pressure. To do this a "cow" or "pig" adaptor can be added to the end of the condenser, or for better results or for very air sensitive compounds a Perkin triangle apparatus can be used.

The Perkin triangle, has means via a series of glass or Teflon taps to allows fractions to be isolated from the rest of the still, without the main body of the distillation being removed from either the vacuum or heat source, and thus can remain in a state of reflux. To do this, the sample is first isolated from the vacuum by means of the taps, the vacuum over the sample is then replaced with an inert gas (such as nitrogen or argon) and can then be stoppered and removed. A fresh collection vessel can then be added to the system, evacuated and linked back into the distillation system via the taps to collect a second fraction, and so on, until all fractions have been collected.

Short Path Distillation

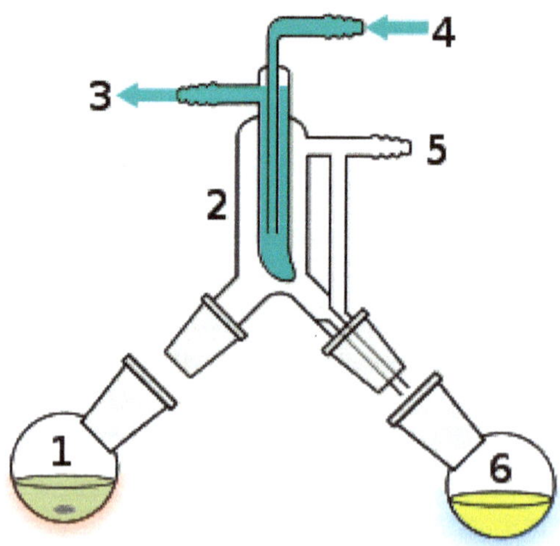

Short path vacuum distillation apparatus with vertical condenser (cold finger), to minimize the distillation path; **1:** Still pot with stirrer bar/anti-bumping granules **2:** Cold finger – bent to direct condensate **3:** Cooling water out **4:** cooling water in **5:** Vacuum/gas inlet **6:** Distillate flask/distillate.

Short path distillation is a distillation technique that involves the distillate travelling a short distance, often only a few centimeters, and is normally done at reduced pressure. A classic example would be a distillation involving the distillate travelling from one glass bulb to another, without the need for a condenser separating the two chambers. This technique is often used for compounds which are unstable at high temperatures or to purify small amounts of compound. The advantage is that the heating temperature can be considerably lower (at reduced pressure) than the boiling point of the liquid at standard pressure, and the distillate only has to travel a short distance before condensing. A short path ensures that little compound is lost on the sides of the apparatus. The Kugelrohr is a kind of a short path distillation apparatus which often contain multiple chambers to collect distillate fractions.

Zone distillation

Zone distillation is a distillation process in long container with partial melting of refined matter in moving liquid zone and condensation of vapor in the solid phase at condensate pulling in cold area. The process is worked in theory. When zone heater is moving from the top to the bottom of the container then solid condensate with irregular impurity distribution is forming. Then most pure part of the condensate may be extracted as product. The process may be iterated many times by moving (without turnover) the received condensate to the bottom part of the container on the place of refined matter. The irregular impurity distribution in the condensate (that is efficiency of purification) increases with number of repetitions of the process. Zone distillation is a distillation analog of zone recrystallization. Impurity distribution in the condensate is described by known equations of zone recrystallization with various numbers of iteration of process – with replacement distribution efficient k of crystallization on separation factor α of distillation.

Other Types

- The process of reactive distillation involves using the reaction vessel as the still. In this process, the product is usually significantly lower-boiling than its reactants. As the product is formed from the reactants, it is vaporized and removed from the reaction mixture. This technique is an example of a continuous vs. a batch process; advantages include less downtime to charge the reaction vessel with starting material, and less workup. Distillation "over a reactant" could be classified as a reactive distillation. It is typically used to remove volatile impurity from the distallation feed. For example, a little lime may be added to remove carbon dioxide from water followed by a second distillation with a little sulfuric acid added to remove traces of ammonia.
- Catalytic distillation is the process by which the reactants are catalyzed while being distilled to continuously separate the products from the reactants. This method is used to assist equilibrium reactions reach completion.
- Pervaporation is a method for the separation of mixtures of liquids by partial vaporization through a non-porous membrane.

- Extractive distillation is defined as distillation in the presence of a miscible, high boiling, relatively non-volatile component, the solvent, that forms no azeotrope with the other components in the mixture.
- Flash evaporation (or partial evaporation) is the partial vaporization that occurs when a saturated liquid stream undergoes a reduction in pressure by passing through a throttling valve or other throttling device. This process is one of the simplest unit operations, being equivalent to a distillation with only one equilibrium stage.
- Codistillation is distillation which is performed on mixtures in which the two compounds are not miscible.

The unit process of evaporation may also be called "distillation":

- In rotary evaporation a vacuum distillation apparatus is used to remove bulk solvents from a sample. Typically the vacuum is generated by a water aspirator or a membrane pump.
- In a kugelrohr a short path distillation apparatus is typically used (generally in combination with a (high) vacuum) to distill high boiling (> 300 °C) compounds. The apparatus consists of an oven in which the compound to be distilled is placed, a receiving portion which is outside of the oven, and a means of rotating the sample. The vacuum is normally generated by using a high vacuum pump.

Other uses:

- Dry distillation or destructive distillation, despite the name, is not truly distillation, but rather a chemical reaction known as pyrolysis in which solid substances are heated in an inert or reducing atmosphere and any volatile fractions, containing high-boiling liquids and products of pyrolysis, are collected. The destructive distillation of wood to give methanol is the root of its common name – *wood alcohol.*
- Freeze distillation is an analogous method of purification using freezing instead of evaporation. It is not truly distillation, but a recrystallization where the product is the mother liquor, and does not produce products equivalent to distillation. This process is used in the production of ice beer and ice wine to increase ethanol and sugar content, respectively. It is also used to produce applejack. Unlike distillation, freeze distillation concentrates poisonous congeners rather than removing them; As a result, many countries prohibit such applejack as a health measure. However, reducing methanol with the absorption of 4A molecular sieve is a practical method for production. Also, distillation by evaporation can separate these since they have different boiling points.

Azeotropic Distillation

Interactions between the components of the solution create properties unique to the solution, as most processes entail nonideal mixtures, where Raoult's law does not hold. Such interactions can result in a constant-boiling **azeotrope** which behaves as if it were a pure compound (i.e., boils at a single temperature instead of a range). At an azeotrope, the solution contains the given component in the same proportion as the vapor, so that evaporation does not change the purity, and distillation does not effect separation. For example, ethyl alcohol and water form an azeotrope of 95.6% at 78.1 °C.

If the azeotrope is not considered sufficiently pure for use, there exist some techniques to break the azeotrope to give a pure distillate. This set of techniques are known as **azeotropic distillation**. Some techniques achieve this by "jumping" over the azeotropic composition (by adding another component to create a new azeotrope, or by varying the pressure). Others work by chemically or physically removing or sequestering the impurity. For example, to purify ethanol beyond 95%, a drying agent (or desiccant, such as potassium carbonate) can be added to convert the soluble water into insoluble water of crystallization. Molecular sieves are often used for this purpose as well.

Immiscible liquids, such as water and toluene, easily form azeotropes. Commonly, these azeotropes are referred to as a low boiling azeotrope because the boiling point of the azeotrope is lower than the boiling point of either pure component. The temperature and composition of the azeotrope is easily predicted from the vapor pressure of the pure components, without use of Raoult's law. The azeotrope is easily broken in a distillation set-up by using a liquid–liquid separator (a decanter) to separate the two liquid layers that are condensed overhead. Only one of the two liquid layers is refluxed to the distillation set up.

High boiling azeotropes, such as a 20 weight percent mixture of hydrochloric acid in water, also exist. As implied by the name, the boiling point of the azeotrope is greater than the boiling point of either pure component.

To break azeotropic distillations and cross distillation boundaries, such as in the DeRosier Problem, it is necessary to increase the composition of the light key in the distillate.

Breaking an Azeotrope with Unidirectional Pressure Manipulation

The boiling points of components in an azeotrope overlap to form a band. By exposing an azeotrope to a vacuum or positive pressure, it's possible to bias the boiling point of one component away from the other by exploiting the differing vapour pressure curves of each; the curves may overlap at the azeotropic point, but are unlikely to be remain identical further along the pressure axis either side of the azeotropic point. When the bias is great enough, the two boiling points no longer overlap and so the azeotropic band disappears.

This method can remove the need to add other chemicals to a distillation, but it has two potential drawbacks.

Under negative pressure, power for a vacuum source is needed and the reduced boiling points of the distillates requires that the condenser be run cooler to prevent distillate vapours being lost to the vacuum source. Increased cooling demands will often require additional energy and possibly new equipment or a change of coolant.

Alternatively, if positive pressures are required, standard glassware can not be used, energy must be used for pressurization and there is a higher chance of side reactions occurring in the distillation, such as decomposition, due to the higher temperatures required to effect boiling.

A unidirectional distillation will rely on a pressure change in one direction, either positive or negative.

Pressure-Swing Distillation

Pressure-swing distillation is essentially the same as the unidirectional distillation used to break azeotropic mixtures, but here both positive and negative pressures may be employed.

This improves the selectivity of the distillation and allows a chemist to optimize distillation by avoiding extremes of pressure and temperature that waste energy. This is particularly important in commercial applications.

One example of the application of pressure-swing distillation is during the industrial purification of ethyl acetate after its catalytic synthesis from ethanol.

Industrial Distillation

Typical industrial distillation towers.

Large scale **industrial distillation** applications include both batch and continuous fractional, vacuum, azeotropic, extractive, and steam distillation. The most widely used industrial applications of continuous, steady-state fractional distillation are in petroleum refineries, petrochemical and chemical plants and natural gas processing plants.

To control and optimize such industrial distillation, a standardized laboratory method, ASTM D86, is established. This test method extends to the atmospheric distillation of petroleum products using a laboratory batch distillation unit to quantitatively determine the boiling range characteristics of petroleum products.

Industrial distillation is typically performed in large, vertical cylindrical columns known as **distillation towers** or **distillation columns** with diameters ranging from about 65 centimeters to 16 meters and heights ranging from about 6 meters to 90 meters or more. When the process feed has a diverse composition, as in distilling crude oil, liquid outlets at intervals up the column allow for the withdrawal of different *fractions* or products having different boiling points or boiling ranges. The "lightest" products (those with the lowest boiling point) exit from the top of the columns and the "heaviest" products (those with the highest boiling point) exit from the bottom of the column and are often called the **bottoms**.

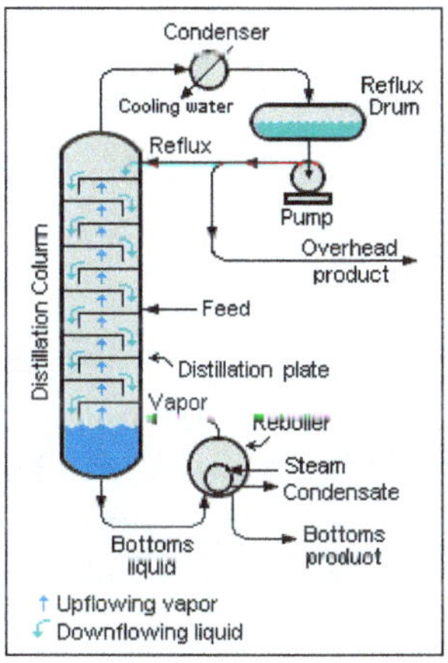

Diagram of a typical industrial distillation tower.

Industrial towers use reflux to achieve a more complete separation of products. Reflux refers to the portion of the condensed overhead liquid product from a distillation or fractionation tower that is returned to the upper part of the tower as shown in the schematic diagram of a typical, large-scale industrial distillation tower. Inside the tower, the downflowing reflux liquid provides cooling and condensation of the upflowing vapors thereby increasing the efficiency of the distillation tower. The more reflux that is provided for a given number of theoretical plates, the better the tower's separation of lower boiling materials from higher boiling materials. Alternatively, the more reflux that is provided for a given desired separation, the fewer the number of theoretical plates required. Chemical engineers must choose what combination of reflux rate and number of plates is both economically and physically feasible for the products purified in the distillation column.

Such industrial fractionating towers are also used in cryogenic air separation, producing liquid oxygen, liquid nitrogen, and high purity argon. Distillation of chlorosilanes also enables the production of high-purity silicon for use as a semiconductor.

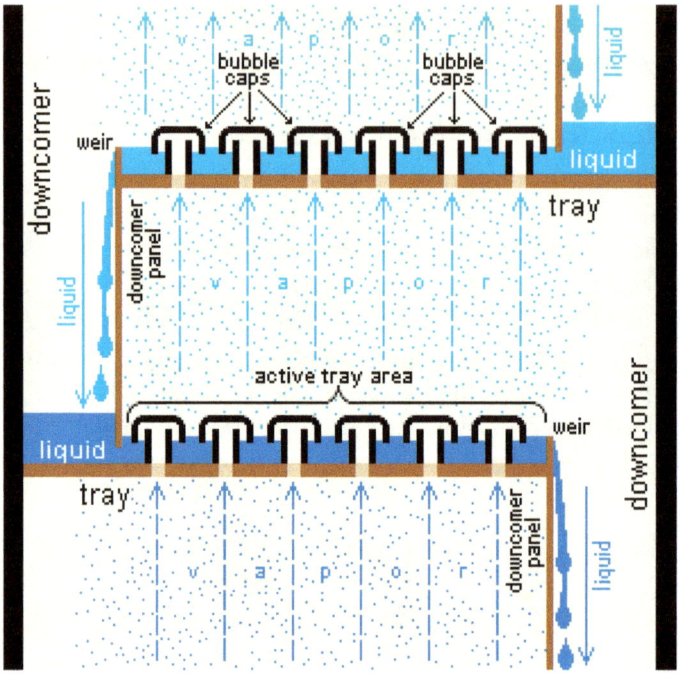

Section of an industrial distillation tower showing detail of trays with bubble caps.

Design and operation of a distillation tower depends on the feed and desired products. Given a simple, binary component feed, analytical methods such as the McCabe–Thiele method or the Fenske equation can be used. For a multi-component feed, simulation models are used both for design and operation. Moreover, the efficiencies of the vapor–liquid contact devices (referred to as "plates" or "trays") used in distillation towers are typically lower than that of a theoretical 100% efficient equilibrium stage. Hence, a distillation tower needs more trays than the number of theoretical vapor–liquid equilibrium stages. A variety of models have been postulated to estimate tray efficiencies.

In modern industrial uses, a packing material is used in the column instead of trays when low pressure drops across the column are required. Other factors that favor packing are: vacuum systems, smaller diameter columns, corrosive systems, systems prone to foaming, systems requiring low liquid holdup, and batch distillation. Conversely, factors that favor plate columns are: presence of solids in feed, high liquid rates, large column diameters, complex columns, columns with wide feed composition variation, columns with a chemical reaction, absorption columns, columns limited by foundation weight tolerance, low liquid rate, large turn-down ratio and those processes subject to process surges.

Large-scale, industrial vacuum distillation column.

This packing material can either be random dumped packing (1–3" wide) such as Raschig rings or structured sheet metal. Liquids tend to wet the surface of the packing and the vapors pass across this wetted surface, where mass transfer takes place. Unlike conventional tray distillation in which every tray represents a separate point of vapor–liquid equilibrium, the vapor–liquid equilibrium curve in a packed column is continuous. However, when modeling packed columns, it is useful to compute a number of "theoretical stages" to denote the separation efficiency of the packed column with respect to more traditional trays. Differently shaped packings have different surface areas and void space between packings. Both of these factors affect packing performance.

Another factor in addition to the packing shape and surface area that affects the performance of random or structured packing is the liquid and vapor distribution entering the packed bed. The number of theoretical stages required to make a given separation is calculated using a specific vapor to liquid ratio. If the liquid and vapor are not evenly distributed across the superficial tower area as it enters the packed bed, the liquid to vapor ratio will not be correct in the packed bed and the required separation will not be achieved. The packing will appear to not be working properly. The height equivalent to a theoretical plate (HETP) will be greater than expected. The problem is not the packing itself but the mal-distribution of the fluids entering the packed bed. Liquid mal-distribution is more frequently the problem than vapor. The design of the liquid distributors used to introduce the feed and reflux to a packed bed is critical to making the packing perform to it maximum efficiency. Methods of evaluating the effectiveness of a liquid distributor to evenly distribute the liquid entering a packed bed can be found in references. Considerable work as been done on this topic by Fractionation Research,

Inc. (commonly known as FRI).

Multi-effect Distillation

The goal of multi-effect distillation is to increase the energy efficiency of the process, for use in desalination, or in some cases one stage in the production of ultrapure water. The number of effects is inversely proportional to the kW·h/m^3 of water recovered figure, and refers to the volume of water recovered per unit of energy compared with single-effect distillation. One effect is roughly 636 kW·h/m^3.

- Multi-stage flash distillation Can achieve more than 20 effects with thermal energy input, as mentioned in the article.
- Vapor compression evaporation Commercial large-scale units can achieve around 72 effects with electrical energy input, according to manufacturers.

There are many other types of multi-effect distillation processes, including one referred to as simply multi-effect distillation (MED), in which multiple chambers, with intervening heat exchangers, are employed.

Distillation in Food Processing

Distilled Beverages

Carbohydrate-containing plant materials are allowed to ferment, producing a dilute solution of ethanol in the process. Spirits such as whiskey and rum are prepared by distilling these dilute solutions of ethanol. Components other than ethanol, including water, esters, and other alcohols, are collected in the condensate, which account for the flavor of the beverage. Some of these beverages are then stored in barrels or other containers to acquire more flavor compounds and characteristic flavors.

Gallery

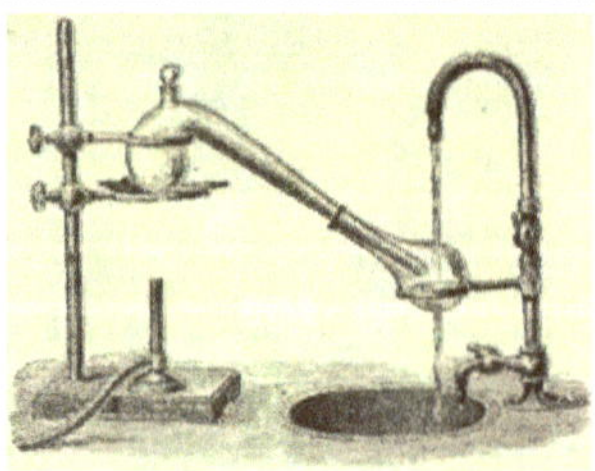	Chemistry in its beginnings used retorts as laboratory equipment exclusively for distillation processes.

A simple set-up to distill dry and oxygen-free toluene.

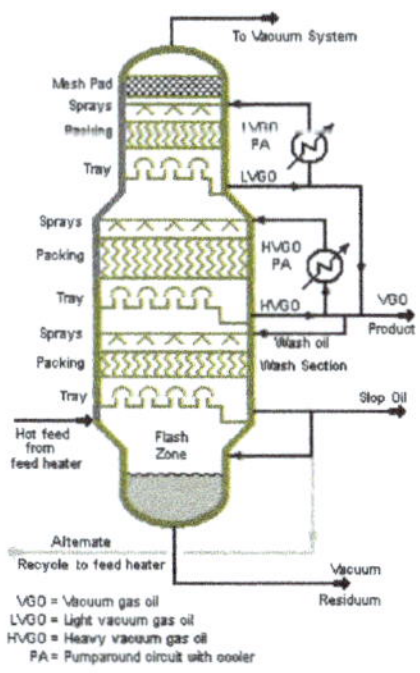

Diagram of an industrial-scale vacuum distillation column as commonly used in oil refineries

A rotary evaporator is able to distill solvents more quickly at lower temperatures through the use of a vacuum.

Distillation using semi-microscale apparatus. The jointless design eliminates the need to fit pieces together. The pear-shaped flask allows the last drop of residue to be removed, compared with a similarly-sized round-bottom flask The small holdup volume prevents losses. A pig is used to channel the various distillates into three receiving flasks. If necessary the distillation can be carried out under vacuum using the vacuum adapter at the pig.

REFERENCES

1. Harwood, Laurence M.; Moody, Christopher J. (1989). Experimental organic chemistry: Principles and Practice (Illustrated ed.). Oxford: Blackwell Scientific Publications. pp. 141–143. ISBN 978-0-632-02017-1.
2. E. Gildemeister; Fr. Hoffman; translated by Edward Kremers (1913). The Volatile Oils. **1**. New York: Wiley. p. 203.
3. Bryan H. Bunch; Alexander Hellemans (2004). The History of Science and Technology. Houghton Mifflin Harcourt. p. 88. ISBN 0-618-22123-9.

4. Marcelin Berthelot Collection des anciens alchimistes grecs (3 vol., Paris, 1887–1888, p.161)

5. Berthelot, M. P. E. M. (1893). The Discovery of Alcohol and Distillation". The Popular Science Monthly. XLIII: 85–94.

6. Forbes, Robert James (1970). A short history of the art of distillation: from the beginnings up to the death of Cellier Blumenthal. BRILL. pp. 57, 89. ISBN 978-90-04-00617-1. Retrieved 29 June 2010.

7. Taylor, F. (1945). The evolution of the still". Annals of Science. **5** *(3): 185.* doi:10.1080/00033794500201451.

8. Haw, Stephen G. (2012). Wine, women and poison". Marco Polo in China. Routledge. pp. 147–148. ISBN 978-1-134-27542-7. The earliest possible period seems to be the Eastern Han dynasty ... the most likely period for the beginning of true distillation of spirits for drinking in China is during the Jin and Southern Song dynasties

9. Sarton, George (1975). Introduction to the history of science. R. E. Krieger Pub. Co. p. 145. ISBN 0-88275-172-7.

10. Holmyard, Eric John (1990). Alchemy. Courier Dover Publications. p. 53. ISBN 0-486-26298-7.

11. Magnum Opus Hermetic Sourceworks Series. Alchemywebsite.com. Retrieved on 2014-04-20.

12. Industrial Engineering Chemistry (1936) p. 677

13. Sealing Technique, accessed 16 November 2006.

14. Traditional Alembic Pot Still, accessed 16 November 2006.

15. D. F. Othmer (1982) Distillation – Some Steps in its Development", in W. F. Furter (ed) A Century of Chemical Engineering ISBN 0-306-40895-3

16. GB 5974, Coffey, A., "Apparatus for Brewing and Distilling", published 5 August 1830, issued 5 February 1831; image

17. Forbes, Robert J. A Short History of the Art of Distillation.

18. US 198699, Solvay, Ernest, "Improvement in the Ammonia-Soda Manufacture", published 2 June 1876, issued 25 December 1877

19. ST07 Separation of liquid–liquid mixtures (solutions), DIDAC by IUPAC

20. Perry, Robert H.; Green, Don W. (1984). Perry's Chemical Engineers' Handbook (6th ed.). McGraw-Hill. ISBN 0-07-049479-7.

21. Fractional Distillation. fandm.edu

22. Spinning Band Distillation. B/R Instrument Corporation (accessed 8 September 2006)

23. Harwood & Moody 1989, pp. 151–153

24. Vogel's 5th ed.

25. Harwood & Moody 1989, p. 150

26. Kravchenko, A. I. (2011). Zone distillation: a new method of refining". Problems

of atomic science and technology (in Russian). **6** *(19): 24–26.*

27. Kravchenko, A. I. (2014). Zone distillation: justification". Problems of atomic science and technology. **1** *(20): 64–65.*
28. Kravchenko, A. I. (2014). Design of advanced processes of zone distillation". Perspectivnye materialy *(in Russian) (7): 68–72.*
29. Study on Method of Decreasing Methanol in Apple Pomace Spirit.
30. Kister, Henry Z. (1992). Distillation Design (1st ed.). McGraw-Hill. ISBN 0-07-034909-6.
31. Seader, J. D.; Henley, Ernest J. (1998). Separation Process Principles. New York: Wiley. ISBN 0-471-58626-9.
32. Energy Institute website page. Resources.schoolscience.co.uk. Retrieved on 2014-04-20.
33. Moore, F., Rukovena, F. (August 1987) Random Packing, Vapor and Liquid Distribution: Liquid and gas distribution in commercial packed towers, Chemical Plants & Processing, Edition Europe, pp. 11–15
34. Spiegel, L (2006). A new method to assess liquid distributor quality". Chemical Engineering and Processing. **45** *(11): 1011. doi:10.1016/j.cep.2006.05.003.*
35. Kunesh, John G.; Lahm, Lawrence; Yanagi, Takashi (1987). Commercial scale experiments that provide insight on packed tower distributors". Industrial & Engineering Chemistry Research. **26** *(9): 1845.* doi:10.1021/ie00069a021.

CHAPTER 6

Anaerobic Digestion

Chapter Contents

Anaerobic digestion is a biological process whereby bacteria break down organic material into more basic compounds without requiring oxygen as a component of the process. These bacteria are believed to have appeared on Earth approximately 3,800,000,000 years ago and were the dominant form of life on the planet before plants appeared. As plant life arose around 3,200,000,000 years ago, anaerobic digestion continued in natural environments where oxygen was absent such as swamps, water-logged soils, and in ground constantly covered by water such as lakes and rivers. The biological processes of anaerobic digestion require that several types of bacteria decompose organic matter in a series of four steps, including hydrolysis, fermentation, acetogenesis, and methanogenesis.

INTRODUCTION

Anaerobic digestion is a collection of processes by which microorganisms break down biodegradable material in the absence of oxygen. The process is used for industrial or domestic purposes to manage waste or to produce fuels. Much of the fermentation used industrially to produce food and drink products, as well as home fermentation, uses anaerobic digestion.

Anaerobic digestion occurs naturally in some soils and in lake and oceanic basin sediments, where it is usually referred to asanaerobic activity". This is the source of marsh gas methane as discovered by Volta in 1776.

The digestion process begins with bacterial hydrolysis of the input materials. Insoluble organic polymers, such as carbohydrates, are broken down to soluble derivatives that become available for other bacteria. Acidogenic bacteria then convert the sugars and amino acids into carbon dioxide, hydrogen, ammonia, and organic acids. These bacteria convert these resulting organic acids into acetic acid, along with additional ammonia, hydrogen, and carbon dioxide. Finally, methanogens convert these products to methane and carbon dioxide. The methanogenic archaea populations play an indispensable role in anaerobic wastewater treatments.

Anaerobic digestion is used as part of the process to treat biodegradable waste and sewage sludge. As part of an integrated waste management system, anaerobic digestion reduces the emission of landfill gas into the atmosphere. Anaerobic digesters can also be fed with purpose-grown energy crops, such as maize.

Anaerobic digestion is widely used as a source of renewable energy. The process produces a biogas, consisting of methane, carbon dioxide and traces of other 'contaminant' gases. This biogas can be used directly as fuel, in combined heat and power gas engines or upgraded to natural gas-quality biomethane. The nutrient-rich digestate also produced can be used as fertilizer.

With the re-use of waste as a resource and new technological approaches that have lowered capital costs, anaerobic digestion has in recent years received increased attention among governments in a number of countries, among these the United Kingdom (2011), Germany and Denmark (2011).

PROCESS

Many microorganisms affect anaerobic digestion, including acetic acid-forming bacteria (acetogens) and methane-forming archaea (methanogens). These organisms promote a number of chemical processes in converting the biomass to biogas.

Gaseous oxygen is excluded from the reactions by physical containment. Anaerobes utilize electron acceptors from sources other than oxygen gas. These acceptors can be the organic material itself or may be supplied by inorganic oxides from within the input material. When the oxygen source in an anaerobic system is derived from the organic material itself, the 'intermediate' end products are primarily alcohols, aldehydes, and organic acids, plus carbon dioxide. In the presence of specialised methanogens, the intermediates are converted to the 'final' end products of methane, carbon dioxide, and trace levels of hydrogen sulfide. In an anaerobic system, the majority of the chemical energy contained within the starting material is released by methanogenic bacteria as methane.

Populations of anaerobic microorganisms typically take a significant period of time to establish themselves to be fully effective. Therefore, common practice is to introduce anaerobic microorganisms from materials with existing populations, a process known asseeding" the digesters, typically accomplished with the addition of sewage sludge or cattle slurry.

Process Stages

The four key stages of anaerobic digestion involve hydrolysis, acidogenesis, acetogenesis and methanogenesis. The overall process can be described by the chemical reaction, where organic material such as glucose is biochemically digested into carbon dioxide (CO_2) and methane (CH_4) by the anaerobic microorganisms.

$$C_6H_{12}O_6 \rightarrow 3CO_2 + 3CH_4$$

- Hydrolysis

In most cases, biomass is made up of large organic polymers. For the bacteria in anaerobic digesters to access the energy potential of the material, these chains must first be broken down into their smaller constituent parts. These constituent parts, or monomers, such as sugars, are readily available to other bacteria. The process of breaking these chains and dissolving the smaller molecules into solution is called hydrolysis. Therefore, hydrolysis of these high-molecular-weight polymeric components is the necessary first step in anaerobic digestion. Through hydrolysis the complex organic molecules are broken down into simple sugars, amino acids, and fatty acids.

Acetate and hydrogen produced in the first stages can be used directly by methanogens. Other molecules, such as volatile fatty acids (VFAs) with a chain length greater than that of acetate must first be catabolised into compounds that can be directly used by methanogens.

- Acidogenesis

The biological process of acidogenesis results in further breakdown of the remaining components by acidogenic (fermentative) bacteria. Here, VFAs are created, along with ammonia, carbon dioxide, and hydrogen sulfide, as well as other byproducts. The process of acidogenesis is similar to the way milk sours.

- Acetogenesis

The third stage of anaerobic digestion is acetogenesis. Here, simple molecules created through the acidogenesis phase are further digested by acetogens to produce largely acetic acid, as well as carbon dioxide and hydrogen.

- Methanogenesis

The terminal stage of anaerobic digestion is the biological process of methanogenesis. Here, methanogens use the intermediate products of the preceding stages and convert them into methane, carbon dioxide, and water. These components make up the majority of the biogas emitted from the system. Methanogenesis is sensitive to both high and low pHs and occurs between pH 6.5 and pH 8. The remaining, indigestible material the microbes cannot use and any dead bacterial remains constitute the digestate.

Configuration

Anaerobic digesters can be designed and engineered to operate using a number of different configurations and can be categorized into batch vs. continuous process mode, mesophilic vs. thermophilic temperature conditions, high vs. low portion of solids, and single stage vs. multistage processes. More initial build money and a larger volume of the batch digester is needed to handle the same amount of waste as a continuous process digester. Higher heat energy is demanded in a thermophilic system compared to a mesophilic system and has a larger gas output capacity and higher methane gas content. For solids content, low will handle up to 15% solid content. Above this level is considered high solids content and can also be known as dry digestion. In a single stage process, one reactor houses the four anaerobic digestion steps. A multistage process utilizes two or more reactors for digestion to separate the methanogenesis and hydrolysis phases.

Batch or Continuous

Anaerobic digestion can be performed as a batch process or a continuous process. In a batch system, biomass is added to the reactor at the start of the process. The reactor is then sealed for the duration of the process. In its simplest form batch processing needs inoculation with already processed material to start the anaerobic digestion. In a typical scenario, biogas production will be formed with a normal distribution pattern over time. Operators can use this fact to determine when they believe the process of digestion of the organic matter has completed. There can be severe odour issues if a batch reactor is opened and emptied before the process is well completed. A more advanced type of batch approach has limited the odour issues by integrating anaerobic digestion with in-vessel composting. In this approach inoculation takes place through the use of recirculated degasified percolate. After anaerobic digestion has completed, the biomass is kept in the reactor which is then used for in-vessel composting before it is opened As the batch digestion is simple and requires less equipment and lower levels of design work, it is typically a cheaper form of digestion. Using more than one batch reactor at a plant can ensure constant production of biogas.

In continuous digestion processes, organic matter is constantly added (continuous complete mixed) or added in stages to the reactor (continuous plug flow; first in – first out). Here, the end products are constantly or periodically removed, resulting in constant production of biogas. A single or multiple digesters in sequence may be used. Examples of this form of anaerobic digestion include continuous stirred-tank reactors, upflow anaerobic sludge blankets, expanded granular sludge beds and internal circulation reactors.

Temperature

The two conventional operational temperature levels for anaerobic digesters determine the species of methanogens in the digesters:

- *Mesophilic* digestion takes place optimally around 30 to 38 °C, or at ambient temperatures between 20 and 45 °C, where mesophiles are the primary microorganism present.
- *Thermophilic* digestion takes place optimally around 49 to 57 °C, or at elevated temperatures up to 70 °C, where thermophiles are the primary microorganisms present.

A limit case has been reached in Bolivia, with anaerobic digestion in temperature working conditions of less than 10 °C. The anaerobic process is very slow, taking more than three times the normal mesophilic time process. In experimental work at University of Alaska Fairbanks, a 1,000 litre digester using psychrophiles harvested from "mud from a frozen lake in Alaska" has produced 200–300 litres of methane per day, about 20 to 30% of the output from digesters in warmer climates. Mesophilic species outnumber thermophiles, and they are also more tolerant to changes in environmental conditions than thermophiles. Mesophilic systems are, therefore, considered to be more stable than thermophilic digestion systems. In contrast, while thermophilic digestion systems are considered less stable, their energy input is higher, with more biogas being removed from the organic matter in an equal amount of time. The increased temperatures facilitate faster reaction rates, and thus faster gas yields. Operation at higher temperatures facilitates greater pathogen reduction of the digestate. In countries where legislation, such as the Animal By-Products Regulations in the European Union, requires digestate to meet certain levels of pathogen reduction there may be a benefit to using thermophilic temperatures instead of mesophilic.

Additional pre-treatment can be used to reduce the necessary retention time to produce biogas. For example, certain processes shred the substrates to increase the surface area or use a thermal pretreatment stage (such as pasteurisation) to significantly enhance the biogas output. The pasteurisation process can also be used to reduce the pathogenic concentration in the digesate leaving the anaerobic digester. Pasteurisation may be achieved by heat treatment combined with maceration of the solids.

Solids Content

In a typical scenario, three different operational parameters are associated with the solids content of the feedstock to the digesters:

- High solids (dry stackable substrate)
- High solids (wet—pumpable substrate)
- Low solids (wet—pumpable substrate)

High solids (dry) digesters are designed to process materials with a solids content between 25 and 40%. Unlike wet digesters that process pumpable slurries, high solids (dry – stackable substrate) digesters are designed to process solid substrates without the addition of water. The primary styles of dry digesters are continuous vertical plug flow and batch tunnel horizontal digesters. Continuous vertical plug flow digesters are upright, cylindrical tanks where feedstock is continuously fed into the top of the digester, and flows downward by gravity during digestion. In batch tunnel digesters, the feedstock is deposited in tunnel-like chambers with a gas-tight door. Neither approach has mixing inside the digester. The amount of pretreatment, such as contaminant removal, depends both upon the nature of the waste streams being processed and the desired quality of the digestate. Size reduction (grinding) is beneficial in continuous vertical systems, as it accelerates digestion, while batch systems avoid grinding and instead require structure (e.g. yard waste) to reduce compaction of the stacked pile. Continuous vertical dry digesters have a smaller footprint due to the shorter effective retention time and vertical design. Wet digesters can be designed to operate in either a high-solids content, with a total suspended solids (TSS) concentration greater than ~20%, or a low-solids concentration less than ~15%.

High solids (wet) digesters process a thick slurry that requires more energy input to move and process the feedstock. The thickness of the material may also lead to associated problems with abrasion. High solids digesters will typically have a lower land requirement due to the lower volumes associated with the moisture. High solids digesters also require correction of conventional performance calculations (e.g. gas production, retention time, kinetics, etc.) originally based on very dilute sewage digestion concepts, since larger fractions of the feedstock mass are potentially convertible to biogas.

Low solids (wet) digesters can transport material through the system using standard pumps that require significantly lower energy input. Low solids digesters require a larger amount of land than high solids due to the increased volumes associated with the increased liquid-to-feedstock ratio of the digesters. There are benefits associated with operation in a liquid environment, as it enables more thorough circulation of materials and contact between the bacteria and their food. This enables the bacteria to more readily access the substances on which they are feeding, and increases the rate of gas production.

Complexity

Digestion systems can be configured with different levels of complexity. In a **single-stage digestion system** (one-stage), all of the biological reactions occur within a single, sealed reactor or holding tank. Using a single stage reduces construction costs, but results in less control of the reactions occurring within the system. Acidogenic bacteria, through the production of acids, reduce the pH of the tank. Methanogenic bacteria, as outlined earlier, operate in a strictly defined pH range. Therefore, the biological reactions of the different species in a single-stage reactor can be in direct competition with each other. Another one-stage reaction system is an anaerobic lagoon. These lagoons are pond-like, earthen basins used for the treatment and long-term storage of manures. Here the anaerobic reactions are contained within the natural anaerobic sludge contained in the pool.

In a **two-stage digestion system** (multistage), different digestion vessels are optimised to bring maximum control over the bacterial communities living within the digesters. Acidogenic bacteria produce organic acids and more quickly grow and reproduce than methanogenic bacteria. Methanogenic bacteria require stable pH and temperature to optimise their performance.

Under typical circumstances, hydrolysis, acetogenesis, and acidogenesis occur within the first reaction vessel. The organic material is then heated to the required operational temperature (either mesophilic or thermophilic) prior to being pumped into a methanogenic reactor. The initial hydrolysis or acidogenesis tanks prior to the methanogenic reactor can provide a buffer to the rate at which feedstock is added. Some European countries require a degree of elevated heat treatment to kill harmful bacteria in the input waste. In this instance, there may be a pasteurisation or sterilisation stage prior to digestion or between the two digestion tanks. Notably, it is not possible to completely isolate the different reaction phases, and often some biogas is produced in the hydrolysis or acidogenesis tanks.

Residence Time

The residence time in a digester varies with the amount and type of feed material, and with the configuration of the digestion system. In a typical two-stage mesophilic digestion, residence time varies between 15 and 40 days, while for a single-stage thermophilic digestion, residence times is normally faster and takes around 14 days. The plug-flow nature of some of these systems will mean the full degradation of the material may not have been realised in this timescale. In this event, digestate exiting the system will be darker in colour and will typically have more odour.

In the case of an upflow anaerobic sludge blanket digestion (UASB), hydraulic residence times can be as short as 1 hour to 1 day, and solid retention times can be up to 90 days. In this manner, a UASB system is able to separate solids and hydraulic retention times with the use of a sludge blanket. Continuous digesters have mechanical or hydraulic devices, depending on the level of solids in the material, to mix the contents, enabling the bacteria and the food to be in contact. They also allow excess material to be continuously extracted to maintain a reasonably constant volume within the digestion tanks.

Inhibition

The anaerobic digestion process can be inhibited by several compounds, affecting one or more of the bacterial groups responsible for the different organic matter degradation steps. The degree of the inhibition depends, among other factors, on the concentration of the inhibitor in the digester. Potential inhibitors are ammonia, sulfide, light metal ions (Na, K, Mg, Ca, Al), heavy metals, some organics (chlorophenols, halogenated aliphatics, N-substituted aromatics, long chain fatty acids), etc.

Left: Farm-based maize silage digester located near Neumünster in Germany, 2007 - the green, inflatable biogas holder is shown on top of the digester. *Right:* Two-stage, low solids, UASB digestion component of a mechanical biological treatment system near Tel Aviv; the process water is seen in balance tank and sequencing batch reactor, 2005.

FEEDSTOCKS

Anaerobic lagoon and generators at the Cal Poly Dairy, United States.

The most important initial issue when considering the application of anaerobic digestion systems is the feedstock to the process. Almost any organic material can be processed with anaerobic digestion; however, if biogas production is the aim, the level of putrescibility is the key factor in its successful application. The more putrescible (digestible) the material, the higher the gas yields possible from the system.

Feedstocks can include biodegradable waste materials, such as waste paper, grass clippings, leftover food, sewage, and animal waste. Woody wastes are the exception, because they are largely unaffected by digestion, as most anaerobes are unable to degrade lignin. Xylophalgeous anaerobes (lignin consumers) or using high temperature pretreatment, such as pyrolysis, can be used to break down the lignin. Anaerobic digesters can also be fed with specially grown energy crops, such as silage, for dedicated biogas production. In Germany and continental Europe, these facilities are referred to asbiogas" plants. A codigestion or cofermentation plant is typically an agricultural anaerobic digester that accepts two or more input materials for simultaneous digestion.

The length of time required for anaerobic digestion depends on the chemical complexity of the material. Material rich in easily digestible sugars breaks down quickly where as intact lignocellulosic material rich in cellulose and hemicellulose polymers can take much longer to break down. Anaerobic microorganisms are generally unable to break down lignin, the recalcitrant aromatic component of biomass.

Anaerobic digesters were originally designed for operation using sewage sludge and manures. Sewage and manure are not, however, the material with the most potential for anaerobic digestion, as the biodegradable material has already had much of the energy content taken out by the animals that produced it. Therefore, many digesters operate with codigestion of two or more types of feedstock. For example, in a farm-based digester that uses dairy manure as the primary feedstock, the gas production may be significantly increased by adding a second feedstock, e.g., grass and corn (typical on-farm feedstock), or various organic byproducts, such as slaughterhouse waste, fats, oils and grease from restaurants, organic household waste, etc. (typical off-site feedstock).

Digesters processing dedicated energy crops can achieve high levels of degradation and biogas production. Slurry-only systems are generally cheaper, but generate far less energy than those using crops, such as maize and grass silage; by using a modest amount of crop material (30%), an anaerobic digestion plant can increase energy output tenfold for only three times the capital cost, relative to a slurry-only system.

Moisture Content

A second consideration related to the feedstock is moisture content. Dryer, stackable substrates, such as food and yard waste, are suitable for digestion in tunnel-like chambers. Tunnel-style systems typically have near-zero wastewater discharge, as well, so this style of system has advantages where the discharge of digester liquids are a liability. The wetter the material, the more suitable it will be to handling with standard pumps instead of energy-intensive concrete pumps and physical means of movement. Also, the wetter the material, the more volume and area it takes up relative to the levels of gas produced. The moisture content of the target feedstock will also affect what type of system is applied to its treatment. To use a high-solids anaerobic digester for dilute feedstocks, bulking agents, such as compost, should be applied to increase the solids content of the input material. Another key consideration is the carbon:nitrogen ratio of the input material. This ratio is the balance of food a microbe requires to grow; the optimal C:N ratio is 20–30:1. Excess N can lead to ammonia inhibition of digestion.

Contamination

The level of contamination of the feedstock material is a key consideration. If the feedstock to the digesters has significant levels of physical contaminants, such as plastic, glass, or metals, then processing to remove the contaminants will be required for the material to be used. If it is not removed, then the digesters can be blocked and will not function efficiently. It is with this understanding that mechanical biological treatment plants are designed. The higher the level of pretreatment a feedstock requires, the more processing machinery will be required, and, hence, the project will have higher capital costs.

After sorting or screening to remove any physical contaminants from the feedstock, the material is often shredded, minced, and mechanically or hydraulically pulped to increase the surface area available to microbes in the digesters and, hence, increase the speed of digestion. The maceration of solids can be achieved by using a chopper pump to transfer the feedstock material into the airtight digester, where anaerobic treatment takes place.

Substrate Composition

Substrate composition is a major factor in determining the methane yield and methane production rates from the digestion of biomass. Techniques to determine the compositional characteristics of the feedstock are available, while parameters such as solids, elemental, and organic analyses are important for digester design and operation.

APPLICATIONS

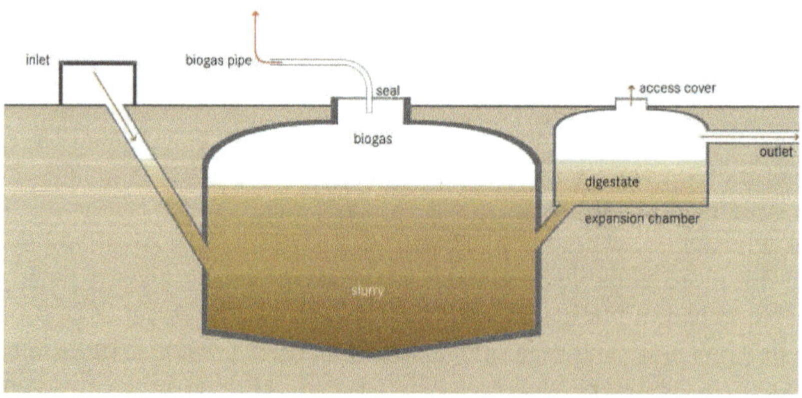

Schematic of an anaerobic digester as part of a sanitation system. It produces a digested slurry (digestate) that can be used as a fertilizer, and biogas that can be used for energy.

Using anaerobic digestion technologies can help to reduce the emission of greenhouse gases in a number of key ways:

- Replacement of fossil fuels
- Reducing or eliminating the energy footprint of waste treatment plants
- Reducing methane emission from landfills
- Displacing industrially produced chemical fertilizers
- Reducing vehicle movements
- Reducing electrical grid transportation losses
- Reducing usage of LP Gas for cooking

Waste and Wastewater Treatment

Anaerobic digestion is particularly suited to organic material, and is commonly used for industrial effluent, wastewater and sewage sludge treatment. Anaerobic digestion, a simple process, can greatly reduce the amount of organic matter which might otherwise be destined to be dumped at sea, dumped in landfills, or burnt in incinerators.

Pressure from environmentally related legislation on solid waste disposal methods in developed countries has increased the application of anaerobic digestion as a process for reducing waste volumes and generating useful byproducts. It may either be used to process the source-separated fraction of municipal waste or alternatively combined with mechanical sorting systems, to process residual mixed municipal waste. These facilities are called mechanical biological treatment plants.

If the putrescible waste processed in anaerobic digesters were disposed of in a landfill, it would break down naturally and often anaerobically. In this case, the gas will eventually escape into the atmosphere. As methane is about 20 times more potent as a greenhouse gas than carbon dioxide, this has significant negative environmental effects.

In countries that collect household waste, the use of local anaerobic digestion facilities can help to reduce the amount of waste that requires transportation to centralized landfill sites or incineration facilities. This reduced burden on transportation reduces carbon emissions from the collection vehicles. If localized anaerobic digestion facilities are embedded within an electrical distribution network, they can help reduce the electrical losses associated with transporting electricity over a national grid.

Power Generation

In developing countries, simple home and farm-based anaerobic digestion systems offer the potential for low-cost energy for cooking and lighting. From 1975, China and India have both had large, government-backed schemes for adaptation of small biogas plants for use in the household for cooking and lighting.At present, projects for anaerobic digestion in the developing world can gain financial support through the United Nations Clean Development Mechanism if they are able to show they provide reduced carbon emissions.

Methane and power produced in anaerobic digestion facilities can be used to replace energy derived from fossil fuels, and hence reduce emissions of greenhouse gases, because the carbon in biodegradable material is part of a carbon cycle. The carbon released into the atmosphere from the combustion of biogas has been removed by plants for them to grow in the recent past, usually within the last decade, but more typically within the last growing season. If the plants are regrown, taking the carbon out of the atmosphere once more, the system will be carbon neutral. In contrast, carbon in fossil fuels has been sequestered in the earth for many millions of years, the combustion of which increases the overall levels of carbon dioxide in the atmosphere.

Biogas from sewage works is sometimes used to run a gas engine to produce electrical power, some or all of which can be used to run the sewage works. Some waste heat from the engine is then used to heat the digester. The waste heat is, in general, enough to heat the digester to the required temperatures. The power potential from sewage works is limited – in the UK, there are about 80 MW total of such generation, with the potential to increase to 150 MW, which is insignificant compared to the average power demand in the UK of about 35,000 MW. The scope for biogas generation from nonsewage waste biological matter – energy crops, food waste, abattoir waste, etc. - is much higher, estimated to be capable of about 3,000 MW. Farm biogas plants using animal waste and energy crops are expected to contribute to reducing CO_2 emissions and strengthen the grid, while providing UK farmers with additional revenues.

Some countries offer incentives in the form of, for example, feed-in tariffs for feeding electricity onto the power grid to subsidize green energy production.

In Oakland, California at the East Bay Municipal Utility District's main wastewater treatment plant (EBMUD), food waste is currently codigested with primary and secondary municipal wastewater solids and other high-strength wastes. Compared to municipal wastewater solids digestion alone, food waste codigestion has many benefits. Anaerobic digestion of food waste pulp from the EBMUD food waste process provides a higher normalized energy benefit, compared to municipal wastewater solids: 730 to 1,300 kWh per dry ton of food waste applied compared to 560 to 940 kWh per dry ton of municipal wastewater solids applied.

Grid Injection

Biogas grid-injection is the injection of biogas into the natural gas grid. The raw biogas has to be previously upgraded to biomethane. This upgrading implies the removal of contaminants such as hydrogen sulphide or siloxanes, as well as the carbon dioxide. Several technologies are available for this purpose, being the most widely implemented the pressure swing adsorption (PSA), water or amine scrubbing (absorption processes) and, in the last years, membrane separation. As an alternative, the electricity and the heat can be used for on-site generation, resulting in a reduction of losses in the transportation of energy. Typical energy losses in natural gas transmission systems range from 1–2%, whereas the current energy losses on a large electrical system range from 5–8%.

In October 2010, Didcot Sewage Works became the first in the UK to produce biomethane gas supplied to the national grid, for use in up to 200 homes in Oxfordshire. By 2017, UK electricity firm Ecotricity plan to have digester fed by locally sourced grass fueling 6000 homes

Vehicle Fuel

After upgrading with the above-mentioned technologies, the biogas (transformed into biomethane) can be used as vehicle fuel in adapted vehicles. This use is very extensive in Sweden, where over 38,600 gas vehicles exist, and 60% of the vehicle gas is biomethane generated in anaerobic digestion plants.

Fertiliser and Soil Conditioner

The solid, fibrous component of the digested material can be used as a soil conditioner to increase the organic content of soils. Digester liquor can be used as a fertiliser to supply vital nutrients to soils instead of chemical fertilisers that require large amounts of energy to produce and transport. The use of manufactured fertilisers is, therefore, more carbon-intensive than the use of anaerobic digester liquor fertiliser. In countries such as Spain, where many soils are organically depleted, the markets for the digested solids can be equally as important as the biogas.

Cooking Gas

By using a bio-digester, which produces the bacteria required for decomposing, cooking gas is generated. The organic garbage like fallen leaves, kitchen waste, food waste etc. are fed into a crusher unit, where the mixture is conflated with a small amount of water. The mixture is then fed into the bio-digester, where the bacteria decomposes it to produce cooking gas. This gas is piped to kitchen stove. A 2 cubic meter bio-digester can produce 2 cubic meter of cooking gas. This is equivalent to 1 kg of LPG. The notable advantage of using a bio-digester is the sludge which is a rich organic manure.

PRODUCTS

The three principal products of anaerobic digestion are biogas, digestate, and water.

Biogas

Typical composition of biogas		
Compound	**Formula**	**%**
Methane	CH4	50–75
Carbon dioxide	CO2	25–50
Nitrogen	N2	0–10

Hydrogen	H2	0–1
Hydrogen sulphide	H2S	0–3
Oxygen	O2	0–0
Source: *www.kolumbus.fi, 2007*		

Biogas is the ultimate waste product of the bacteria feeding off the input biodegradable feedstock (the methanogenesis stage of anaerobic digestion is performed by archaea (a micro-organism on a distinctly different branch of the phylogenetic tree of life to bacteria), and is mostly methane and carbon dioxide, with a small amount hydrogen and trace hydrogen sulfide. (As-produced, biogas also contains water vapor, with the fractional water vapor volume a function of biogas temperature). Most of the biogas is produced during the middle of the digestion, after the bacterial population has grown, and tapers off as the putrescible material is exhausted. The gas is normally stored on top of the digester in an inflatable gas bubble or extracted and stored next to the facility in a gas holder.

The methane in biogas can be burned to produce both heat and electricity, usually with a reciprocating engine or microturbine often in a cogeneration arrangement where the electricity and waste heat generated are used to warm the digesters or to heat buildings. Excess electricity can be sold to suppliers or put into the local grid. Electricity produced by anaerobic digesters is considered to be renewable energy and may attract subsidies. Biogas does not contribute to increasing atmospheric carbon dioxide concentrations because the gas is not released directly into the atmosphere and the carbon dioxide comes from an organic source with a short carbon cycle.

Biogas may require treatment or 'scrubbing' to refine it for use as a fuel. Hydrogen sulfide, a toxic product formed from sulfates in the feedstock, is released as a trace component of the biogas. National environmental enforcement agencies, such as the U.S. Environmental Protection Agency or the English and Welsh Environment Agency, put strict limits on the levels of gases containing hydrogen sulfide, and, if the levels of hydrogen sulfide in the gas are high, gas scrubbing and cleaning equipment (such as amine gas treating) will be needed to process the biogas to within regionally accepted levels. Alternatively, the addition of ferrous chloride $FeCl_2$ to the digestion tanks inhibits hydrogen sulfide production.

Volatile siloxanes can also contaminate the biogas; such compounds are frequently found in household waste and wastewater. In digestion facilities accepting these materials as a component of the feedstock, low-molecular-weight siloxanes volatilise into biogas. When this gas is combusted in a gas engine, turbine, or boiler, siloxanes are converted into silicon dioxide (SiO_2), which deposits internally in the machine, increasing wear and tear. Practical and cost-effective technologies to remove siloxanes and other biogas contaminants are available at the present time. In certain applications, *in situ* treatment can be used to increase the methane purity by reducing the offgas carbon dioxide content, purging the majority of it in a secondary reactor.

In countries such as Switzerland, Germany, and Sweden, the methane in the biogas may be compressed for it to be used as a vehicle transportation fuel or input directly into the gas mains. In countries where the driver for the use of anaerobic digestion are renewable electricity subsidies, this route of treatment is less likely, as energy is required in this processing stage and reduces the overall levels available to sell.

Biogas holder with lightning protection rods and backup gas flare

Digestate

Digestate is the solid remnants of the original input material to the digesters that the microbes cannot use. It also consists of the mineralised remains of the dead bacteria from within the digesters. Digestate can come in three forms: fibrous, liquor, or a sludge-based combination of the two fractions. In two-stage systems, different forms of digestate come from different digestion tanks. In single-stage digestion systems, the two fractions will be combined and, if desired, separated by further processing.

Acidogenic anaerobic digestate.

The second byproduct (acidogenic digestate) is a stable, organic material consisting largely of lignin and cellulose, but also of a variety of mineral components in a matrix of dead bacterial cells; some plastic may be present. The material resembles domestic compost and can be used as such or to make low-grade building products, such as fibreboard. The solid digestate can also be used as feedstock for ethanol production.

The third byproduct is a liquid (methanogenic digestate) rich in nutrients, which can be used as a fertiliser, depending on the quality of the material being digested. Levels of potentially toxic elements (PTEs) should be chemically assessed. This will depend upon the quality of the original feedstock. In the case of most clean and source-separated biodegradable waste streams, the levels of PTEs will be low. In the case of wastes originating from industry, the levels of PTEs may be higher and will need to be taken into consideration when determining a suitable end use for the material.

Digestate typically contains elements, such as lignin, that cannot be broken down by the anaerobic microorganisms. Also, the digestate may contain ammonia that is phytotoxic, and may hamper the growth of plants if it is used as a soil-improving material. For these two reasons, a maturation or composting stage may be employed after digestion. Lignin and other materials are available for degradation by aerobic microorganisms, such as fungi, helping reduce the overall volume of the material for transport. During this maturation, the ammonia will be oxidized into nitrates, improving the fertility of the material and making it more suitable as a soil improver. Large composting stages are typically used by dry anaerobic digestion technologies.

Wastewater

The final output from anaerobic digestion systems is water, which originates both from the moisture content of the original waste that was treated and water produced during the microbial reactions in the digestion systems. This water may be released from the dewatering of the digestate or may be implicitly separate from the digestate.

The wastewater exiting the anaerobic digestion facility will typically have elevated levels of biochemical oxygen demand (BOD) and chemical oxygen demand (COD). These measures of the reactivity of the effluent indicate an ability to pollute. Some of this material is termed 'hard COD', meaning it cannot be accessed by the anaerobic bacteria for conversion into biogas. If this effluent were put directly into watercourses, it would negatively affect them by causing eutrophication. As such, further treatment of the wastewater is often required. This treatment will typically be an oxidation stage wherein air is passed through the water in a sequencing batch reactors or reverse osmosis unit.

HISTORY

Gas street lamp.

Reported scientific interest in the manufacturing of gas produced by the natural decomposition of organic matter dates from the 17th century, when Robert Boyle (1627-1691) and Stephen Hales (1677-1761) noted that disturbing the sediment of streams and lakes released flammable gas. In 1808 Sir Humphry Davy proved the presence of methane in the gases produced by cattle manure.In 1859 a leper colony in Bombay in India built the first anaerobic digester. In 1895, the technology was developed in Exeter, England, where a septic tank was used to generate gas for the sewer gas destructor lamp, a type of gas lighting. Also in England, in 1904, the first dual-purpose tank for both sedimentation and sludge treatment was installed in Hampton, London. In 1907, in Germany, a patent was issued for the Imhoff tank, an early form of digester.

Research on anaerobic digestion began in earnest in the 1930s.

Anaerobic Organism

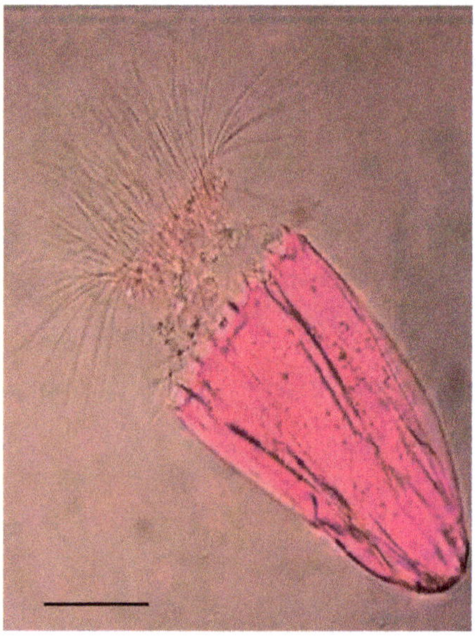

Spinoloricus nov. sp., a metazoan that metabolises with hydrogen, lacking mitochondria and instead using hydrogenosomes.

An **anaerobic organism** or **anaerobe** is any organism that does not require oxygen for growth. It may react negatively or even die if oxygen is present. (In contrast, an aerobic organism (aerobe) is an organism that can survive and grow in an oxygenated environment.)

An anaerobic organism may be unicellular (e.g. protozoans, bacteria) or multicellular. For practical purposes, there are three categories of anaerobe: obligate anaerobes, which are harmed by the presence of oxygen; aerotolerant organisms, which cannot use oxygen for growth but tolerate its presence; and facultative anaerobes, which can grow without oxygen but use oxygen if it is present.

First Observation

In his letter of 14 June 1680 to The Royal Society, Antonie van Leeuwenhoek described an experiment he carried out by filling two identical glass tubes about halfway with crushed pepper powder, to which some clean rain water was added. Van Leeuwenhoek sealed one of the glass tubes by using a flame and left the other glass tube open. Several days later, he discovered in the open glass tube 'a great many very little animalcules, of divers sort having its own particular motion.' Not expecting to see any life in the sealed glass tube, Van Leeuwenhoek saw to his surprise 'a kind of living animalcules that were round and bigger than the biggest sort that I have said were in the other water.'

The conditions in the sealed tube had become quite anaerobic owing to consumption of oxygen by aerobic microorganisms.

In 1913 Martinus Beijerinck repeated Van Leeuwenhoek's experiment and identified Clostridium butyricum as a prominent anaerobic bacterium in the sealed pepper infusion tube liquid. Beijerinck commented:

'We thus come to the remarkable conclusion that, beyond doubt, Van Leeuwenhoek in his experiment with the fully closed tube had cultivated and seen genuine anaerobic bacteria, which would happen again only after 200 years, namely about 1862 by Pasteur. That Leeuwenhoek, one hundred years before the discovery of oxygen and the composition of air, was not aware of the meaning of his observations is understandable. But the fact that in the closed tube he observed an increased gas pressure caused by fermentative bacteria and in addition saw the bacteria, prove in any case that he not only was a good observer, but also was able to design an experiment from which a conclusion could be drawn.'

Classification

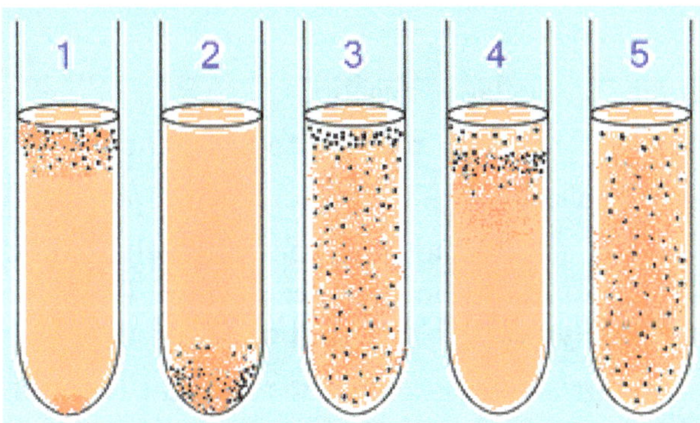

Aerobic and anaerobic bacteria can be identified by growing them in test tubes of thioglycollate broth:

- Obligate aerobes need oxygen because they cannot ferment or respire anaerobically. They gather at the top of the tube where the oxygen concentration is highest.
- Obligate anaerobes are poisoned by oxygen, so they gather at the bottom of the tube where the oxygen concentration is lowest.
- Facultative anaerobes can grow with or without oxygen because they can metabolise energy aerobically or anaerobically. They gather mostly at the top because aerobic respiration generates more ATP than either fermentation or anaerobic respiration.
- Microaerophiles need oxygen because they cannot ferment or respire anaerobically. However, they are poisoned by high concentrations of oxygen. They gather in the upper part of the test tube but not the very top.

- Aerotolerant organisms do not require oxygen as they metabolise energy anaerobically. Unlike obligate anaerobes however, they are not poisoned by oxygen. They can be found evenly spread throughout the test tube.

For practical purposes, there are three categories of anaerobe:

- **Obligate anaerobes**, which are harmed by the presence of oxygen.
- **Aerotolerant organisms**, which cannot use oxygen for growth, but tolerate its presence.
- **Facultative anaerobes**, which can grow without oxygen but use oxygen if it is present.

Energy Metabolism

Some obligate anaerobes use fermentation, while others use anaerobic respiration. Aerotolerant organisms are strictly fermentative. In the presence of oxygen, facultative anaerobes use aerobic respiration; without oxygen, some of them ferment; some use anaerobic respiration.

Fermentation

There are many anaerobic fermentative reactions.

Fermentative anaerobic organisms mostly use the lactic acid fermentation pathway:

$C_6H_{12}O_6$ + 2 ADP + 2 phosphate → 2 lactic acid + 2 ATP

The energy released in this equation is approximately 150 kJ per mol, which is conserved in regenerating two ATP from ADP per glucose. This is only 5% of the energy per sugar molecule that the typical aerobic reaction generates.

Plants and fungi (e.g., yeasts) in general use alcohol (ethanol) fermentation when oxygen becomes limiting:

$C_6H_{12}O_6$ (glucose) + 2 ADP + 2 phosphate → 2 C_2H_5OH + 2 $CO_2\uparrow$ + 2 ATP

The energy released is about 180 kJ per mol, which is conserved in regenerating two ATP from ADP per glucose.

Anaerobic bacteria and archaea use these and many other fermentative pathways, e.g., propionic acid fermentation, butyric acid fermentation, solvent fermentation, mixed acid fermentation, butanediol fermentation, Stickland fermentation, acetogenesis, or methanogenesis.

Culturing Anaerobes

Since normal microbial culturing occurs in atmospheric air, which is an aerobic environment, the culturing of anaerobes poses a problem. Therefore, a number of techniques are employed by microbiologists when culturing anaerobic organisms, for example, handling the bacteria in a glovebox filled with nitrogen or the use of other specially sealed containers, or techniques such as injection of the bacteria into a dicot plant, which is an environment with limited oxygen. The GasPak System is an isolated container that achieves an anaerobic environment by the reaction of water with sodium borohydride and sodium bicarbonate tablets to produce hydrogen gas and carbon dioxide. Hydrogen then reacts with oxygen gas on a palladium catalyst to produce more water, thereby removing oxygen gas. The issue with the Gaspak method is that an adverse reaction can take place where the bacteria may die, which is why a thioglycollate medium should be used. The thioglycollate supplies a medium mimicking that of a dicot, thus providing not only an anaerobic environment but all the nutrients needed for the bacteria to thrive.

Multicellularity

Complex multicellular life that does not need oxygen is said to be rare, however there are examples of such organisms.

At least three species have been discovered in the hypersaline anoxic L'Atalante basin at the bottom of the Mediterranean Sea in 2010, metabolizing with hydrogen, lacking mitochondria and instead using hydrogenosomes.

Some organisms metabolise primarily using glycogen, for example the Nereid (worm) s and some polychaetes, or the juvenile *Trichinella spiralis* (pork worm) parasites.)

Differing Terms and Definitions

Ground source heating and cooling

Some confusion exists with regard to the terminology of heat pumps and the use of the term "*geothermal*". "*Geothermal*" derives from the Greek and means "*Earth heat*" - which geologists and many laymen understand as describing hot rocks, volcanic activity or heat derived from deep within the earth. Though some confusion arises when the term "*geothermal*" is also used to apply to temperatures within the first 100 metres of the surface, this is "*Earth heat*" all the same, though it is largely influenced by stored energy from the sun.

The heat pump was described by Lord Kelvin in 1853 and developed by Peter Ritter von Rittinger in 1855. After experimenting with a freezer, Robert C. Webber built the first direct exchange ground-source heat pump in the late 1940s. The first successful commercial project was installed in the Commonwealth Building (Portland, Oregon) in 1948, and has been designated a National Historic Mechanical Engineering Landmark by ASME. The technology became popular in Sweden in the 1970s, and has been growing slowly in worldwide acceptance since then. Open loop systems dominated the market until the development of polybutylene pipe in 1979 made closed loop systems economically viable. As of 2004, there are over a million units installed worldwide providing 12 GW of thermal capacity. Each year, about 80,000 units are installed in the US (geothermal energy is used in all 50 U.S. states today, with great potential for near-term market growth and savings) and 27,000 in Sweden. In Finland, a geothermal heat pump was the most common heating system choice for new detached houses between 2006 and 2011 with market share exceeding 40%.

Ground Heat Exchanger

Loop field for a 12-ton system (unusually large for most residential applications).

Heat pumps provide winter heating by extracting heat from a source and transferring it into a building. Heat can be extracted from any source, no matter how cold, but a warmer source allows higher efficiency. A ground source heat pump uses the top layer of the earth's crust as a source of heat, thus taking advantage of its seasonally moderated temperature.

In the summer, the process can be reversed so the heat pump extracts heat from the building and transfers it to the ground. Transferring heat to a cooler space takes less energy, so the cooling efficiency of the heat pump gains benefits from the lower ground temperature.

Ground source heat pumps employ a heat exchanger in contact with the ground or groundwater to extract or dissipate heat. This component accounts for anywhere from a fifth to half of the total system cost, and would be the most cumbersome part to repair or replace. Correctly sizing this component is necessary to assure long-term performance: the energy efficiency of the system improves with roughly 4% for every degree Celsius that is won through correct sizing, and the underground temperature balance must be maintained through proper design of the whole system. Incorrect design can result in the system freezing after a number of years or very inefficient system performance; thus accurate system design is critical to a successful system

Shallow 3–8-foot (0.91–2.44 m) horizontal heat exchangers experience seasonal temperature cycles due to solar gains and transmission losses to ambient air at ground level. These temperature cycles lag behind the seasons because of thermal inertia, so the heat exchanger will harvest heat deposited by the sun several months earlier, while being weighed down in late winter and spring, due to accumulated winter cold. Deep vertical systems 100–500 feet (30–152 m) deep rely on migration of heat from surrounding geology, unless they are recharged annually by solar recharge of the ground or exhaust heat from air conditioning systems.

Several major design options are available for these, which are classified by fluid and layout. Direct exchange systems circulate refrigerant underground, closed loop systems use a mixture of anti-freeze and water, and open loop systems use natural groundwater.

Direct Exchange (DX)

The direct exchange geothermal heat pump (DX) is the oldest type of geothermal heat pump technology. The ground-coupling is achieved through a single loop, circulating refrigerant, in direct thermal contact with the ground (as opposed to a combination of a refrigerant loop and a water loop). The refrigerant leaves the heat pump cabinet, circulates through a loop of copper tube buried underground, and exchanges heat with the ground before returning to the pump. The name "direct exchange" refers to heat transfer between the refrigerant loop and the ground without the use of an intermediate fluid. There is no direct interaction between the fluid and the earth; only heat transfer through the pipe wall. Direct exchange heat pumps are not to be confused with "water-source heat pumps" or "water loop heat pumps" since there is no water in the ground loop. ASHRAE defines the term ground-coupled heat pump to encompass closed loop and direct exchange systems, while excluding open loops.

Direct exchange geothermal system.

Direct exchange systems are more efficient and have potentially lower installation costs than closed loop water systems. Copper's high thermal conductivity contributes to the higher efficiency of the system, but heat flow is predominantly limited by the thermal conductivity of the ground, not the pipe. The main reasons for the higher efficiency are the elimination of the water pump (which uses electricity), the elimination of the water-to-refrigerant heat exchanger (which is a source of heat losses), and most importantly, the latent heat phase change of the refrigerant in the ground itself.

However, in case of leakage there is virtually no risk of contaminating the ground or the ground water. Contrary to water-source geothermal systems, direct exchange systems do not contain antifreeze. So, in case of a refrigerant leakage, the refrigerant currently used in most systems - R-410A – would immediately vaporize and seek the atmosphere. This is due to the low boiling point of R-410A: -60.5 °F. R-410A refrigerant replaces larger volumes of antifreeze mixtures used in water-source geothermal systems and presents no threat to aquifers or to the ground itself.

While they require more refrigerant and their tubing is more expensive per foot, a direct exchange earth loop is shorter than a closed water loop for a given capacity. A direct exchange system requires only 15 to 40% of the length of tubing and half the diameter of drilled holes, and the drilling or excavation costs are therefore lower. Refrigerant loops are less tolerant of leaks than water loops because gas can leak out through smaller imperfections. This dictates the use of brazed copper tubing, even though the pressures are similar to water loops. The copper loop must be protected from corrosion in acidic soil through the use of a sacrificial anode or other cathodic protection.

The U.S. Environmental Protection Agency conducted field monitoring of a direct geoexchange heat pump water heating system in a commercial application. The EPA reported that the system saved 75% of the electrical energy that would have been required by an electrical resistance water heating unit. According to the EPA, if the system is operated to capacity, it can avoid the emission of up to 7,100 pounds of CO_2 and 15 pounds of NO_x each year per ton of compressor capacity (or 42,600 lbs. of CO_2 and 90 lbs. of NO_x for a typical 6 ton system).

In Northern climates, although the earth temperature is cooler, so is the incoming water temperature, which enables the high efficiency systems to replace more energy than would otherwise be required of electric or fossil fuel fired systems. Any temperature above -40 °F is sufficient to evaporate the refrigerant, and the direct exchange system can harvest energy through ice.

In extremely hot climates with dry soil, the addition of an auxiliary cooling module as a second condenser in line between the compressor and the earth loops increases efficiency and can further reduce the amount of earth loop to be installed.

Closed Loop

Most installed systems have two loops on the ground side: the primary refrigerant loop is contained in the appliance cabinet where it exchanges heat with a secondary water loop that is buried underground. The secondary loop is typically made of high-density polyethylene pipe and contains a mixture of water and anti-freeze (propylene glycol, denatured alcohol or methanol). Monopropylene glycol has the least damaging potential when it might leak into the ground, and is therefore the only allowed anti-freeze in ground sources in an increasing number of European countries. After leaving the internal heat exchanger, the water flows through the secondary loop outside the building to exchange heat with the ground before returning. The secondary loop is placed below the frost line where the temperature is more stable, or preferably submerged in a body of water if available. Systems in wet ground or in water are generally more efficient than drier ground loops since water conducts and stores heat better than solids in sand or soil. If the ground is naturally dry, soaker hoses may be buried with the ground loop to keep it wet.

Closed loop systems need a heat exchanger between the refrigerant loop and the water loop, and pumps in both loops. Some manufacturers have a separate ground loop fluid pump pack, while some integrate the pumping and valving within the heat pump. Expansion tanks and pressure relief valves may be installed on the heated fluid side. Closed loop systems have lower efficiency than direct exchange systems, so they require longer and larger pipe to be placed in the ground, increasing excavation costs.

Closed loop tubing can be installed horizontally as a loop field in trenches or vertically as a series of long U-shapes in wells (see below). The size of the loop field depends on the soil type and moisture content, the average ground temperature and the heat loss and or gain characteristics of the building being conditioned. A rough approximation of the initial soil temperature is the average daily temperature for the region.

Vertical

Drilling of a borehole for residential heating.

A vertical closed loop field is composed of pipes that run vertically in the ground. A hole is bored in the ground, typically 50 to 400 feet (15–122 m) deep. Pipe pairs in the hole are joined with a U-shaped cross connector at the bottom of the hole. The borehole is commonly filled with a bentonite grout surrounding the pipe to provide a thermal connection to the surrounding soil or rock to improve the heat transfer. Thermally enhanced grouts are available to improve this heat transfer. Grout also protects the ground water from contamination, and prevents artesian wells from flooding the property. Vertical loop fields are typically used when there is a limited area of land available. Bore holes are spaced at least 5–6 m apart and the depth depends on ground and building characteristics. For illustration, a detached house needing 10 kW (3 ton) of heating capacity might need three boreholes 80 to 110 m (260 to 360 ft) deep. (A ton of heat is 12,000 British thermal units per hour (BTU/h) or 3.5 kilowatts.) During the cooling season, the local temperature rise in the bore field is influenced most by the moisture travel in the soil. Reliable heat transfer models have been developed through sample bore holes as well as other tests.

Horizontal

A three-ton slinky loop prior to being covered with soil. The three slinky loops are running out horizontally with three straight lines returning the end of the slinky coil to the heat pump.

A horizontal closed loop field is composed of pipes that run horizontally in the ground. A long horizontal trench, deeper than the frost line, is dug and U-shaped or slinky coils are placed horizontally inside the same trench. Excavation for shallow horizontal loop fields is about half the cost of vertical drilling, so this is the most common layout used wherever there is adequate land available. For illustration, a detached house needing 10 kW (3 ton) of heating capacity might need three loops 120 to 180 m (390 to 590 ft) long of NPS 3/4 (DN 20) or NPS 1.25 (DN 32) polyethylene tubing at a depth of 1 to 2 m (3.3 to 6.6 ft).

The depth at which the loops are placed significantly influences the energy consumption of the heat pump in two opposite ways: shallow loops tend to indirectly absorb more heat from the sun, which is helpful, especially when the ground is still cold after a long winter. On the other hand, shallow loops are also cooled down much more readily by weather changes, especially during long cold winters, when heating demand peaks. Often, the second effect is much greater than the first one, leading to higher costs of operation for the more shallow ground loops. This problem can be reduced by increasing both the depth and the length of piping, thereby significantly increasing costs of installation. However, such expenses might be deemed feasible, as they may result in lower operating costs. Recent studies show that utilization of a non-homogeneous soil profile with a layer of low conductive material above the ground pipes can help mitigate the adverse effects of shallow pipe burial depth. The intermediate blanket with lower conductivity than the surrounding soil profile demonstrated the potential to increase the energy extraction rates from the ground to as high as 17% for a cold climate and about 5-6% for a relatively moderate climate.

A slinky (also called coiled) closed loop field is a type of horizontal closed loop where the pipes overlay each other (not a recommended method). The easiest way of picturing a slinky field is to imagine holding a slinky on the top and bottom with your hands and then moving your hands in opposite directions. A slinky loop field is used if there is not adequate room for a true horizontal system, but it still allows for an easy installation. Rather than using straight pipe, slinky coils use overlapped loops of piping laid out horizontally along the bottom of a wide trench. Depending on soil, climate and the heat pump's run fraction, slinky coil trenches can be up to two thirds shorter than traditional horizontal loop trenches. Slinky coil ground loops are essentially a more economical and space efficient version of a horizontal ground loop.

Radial or Directional Drilling

As an alternative to trenching, loops may be laid by mini horizontal directional drilling (mini-HDD). This technique can lay piping under yards, driveways, gardens or other structures without disturbing them, with a cost between those of trenching and vertical drilling. This system also differs from horizontal & vertical drilling as the loops are installed from one central chamber, further reducing the ground space needed. Radial drilling is often installed retroactively (after the property has been built) due to the small nature of the equipment used and the ability to bore beneath existing constructions.

Pond

12-ton pond loop system being sunk to the bottom of a pond.

A closed pond loop is not common because it depends on proximity to a body of water, where an open loop system is usually preferable. A pond loop may be advantageous where poor water quality precludes an open loop, or where the system heat load is small. A pond loop consists of coils of pipe similar to a slinky loop attached to a frame and located at the bottom of an appropriately sized pond or water source.

Open Loop

In an open loop system (also called a groundwater heat pump), the secondary loop pumps natural water from a well or body of water into a heat exchanger inside the heat pump. ASHRAE calls open loop systems *groundwater heat pumps* or *surface water heat pumps*, depending on the source. Heat is either extracted or added by the primary refrigerant loop, and the water is returned to a separate injection well, irrigation trench, tile field or body of water. The supply and return lines must be placed far enough apart to ensure thermal recharge of the source. Since the water chemistry is not controlled, the appliance may need to be protected from corrosion by using different metals in the heat exchanger and pump. Limescale may foul the system over time and require periodic acid cleaning. This is much more of a problem with cooling systems than heating systems. Also, as fouling decreases the flow of natural water, it becomes difficult for the heat pump to exchange building heat with the groundwater. If the water contains high levels of salt, minerals, iron bacteria or hydrogen sulfide, a closed loop system is usually preferable.

Deep lake water cooling uses a similar process with an open loop for air conditioning and cooling. Open loop systems using ground water are usually more efficient than closed systems because they are better coupled with ground temperatures. Closed loop systems, in comparison, have to transfer heat across extra layers of pipe wall and dirt.

A growing number of jurisdictions have outlawed open-loop systems that drain to the surface because these may drain aquifers or contaminate wells. This forces the use of more environmentally sound injection wells or a closed loop system.

Standing Column Well

A standing column well system is a specialized type of open loop system. Water is drawn from the bottom of a deep rock well, passed through a heat pump, and returned to the top of the well, where traveling downwards it exchanges heat with the surrounding bedrock. The choice of a standing column well system is often dictated where there is near-surface bedrock and limited surface area is available. A standing column is typically not suitable in locations where the geology is mostly clay, silt, or sand. If bedrock is deeper than 200 feet (61 m) from the surface, the cost of casing to seal off the overburden may become prohibitive.

A multiple standing column well system can support a large structure in an urban or rural application. The standing column well method is also popular in residential and small commercial applications. There are many successful applications of varying sizes and well quantities in the many boroughs of New York City, and is also the most common application in the New England states. This type of ground source system has some heat storage benefits, where heat is rejected from the building and the temperature of the well is raised, within reason, during the summer cooling months which can then be harvested for heating in the winter months, thereby increasing the efficiency of the heat pump system. As with closed loop systems, sizing of the standing column system is critical in reference to the heat loss and gain of the existing building. As the heat exchange is actually with the bedrock, using water as the transfer medium, a large amount of production capacity (water flow from the well) is not required for a standing column system to work. However, if there is adequate water production, then the thermal capacity of the well system can be enhanced by discharging a small percentage of system flow during the peak Summer and Winter months.

Since this is essentially a water pumping system, standing column well design requires critical considerations to obtain peak operating efficiency. Should a standing column well design be misapplied, leaving out critical shut-off valves for example, the result could be an extreme loss in efficiency and thereby cause operational cost to be higher than anticipated.

Building Distribution

Liquid-to-air heat pump.

The heat pump is the central unit that becomes the heating and cooling plant for the building. Some models may cover space heating, space cooling, (space heating via conditioned air, hydronic systems and / or radiant heating systems), domestic or pool water preheat (via the desuperheater function), demand hot water, and driveway ice melting all within one appliance with a variety of options with respect to controls, staging and zone control. The heat may be carried to its end use by circulating water or forced air. Almost all types of heat pumps are produced for commercial and residential applications.

Liquid-to-air heat pumps (also called *water-to-air*) output forced air, and are most commonly used to replace legacy forced air furnaces and central air conditioning systems. There are variations that allow for split systems, high-velocity systems, and ductless systems. Heat pumps cannot achieve as high a fluid temperature as a conventional furnace, so they require a higher volume flow rate of air to compensate. When retrofitting a residence, the existing duct work may have to be enlarged to reduce the noise from the higher air flow.

Liquid-to-water heat pump.

Liquid-to-water heat pumps (also called *water-to-water*) are hydronic systems that use water to carry heating or cooling through the building. Systems such as radiant underfloor heating, baseboard radiators, conventional cast iron radiators would use a liquid-to-water heat pump. These heat pumps are preferred for pool heating or domestic hot water pre-heat. Heat pumps can only heat water to about 50 °C (122 °F) efficiently, whereas a boiler normally reaches 65–95 °C (149–203 °F). Legacy radiators designed for these higher temperatures may have to be doubled in numbers when retrofitting a home. A hot water tank will still be needed to raise water temperatures above the heat pump's maximum, but pre-heating will save 25–50% of hot water costs.

Ground source heat pumps are especially well matched to underfloor heating and baseboard radiator systems which only require warm temperatures 40 °C (104 °F) to work well. Thus they are ideal for open plan offices. Using large surfaces such as floors, as opposed to radiators, distributes the heat more uniformly and allows for a lower water temperature. Wood or carpet floor coverings dampen this effect because the thermal transfer efficiency of these materials is lower than that of masonry floors (tile, concrete). Underfloor piping, ceiling or wall radiators can also be used for cooling in dry climates, although the temperature of the circulating water must be above the dew point to ensure that atmospheric humidity does not condense on the radiator.

Combination heat pumps are available that can produce forced air and circulating water simultaneously and individually. These systems are largely being used for houses that have a combination of air and liquid conditioning needs, for example central air conditioning and pool heating.

Seasonal Thermal Storage

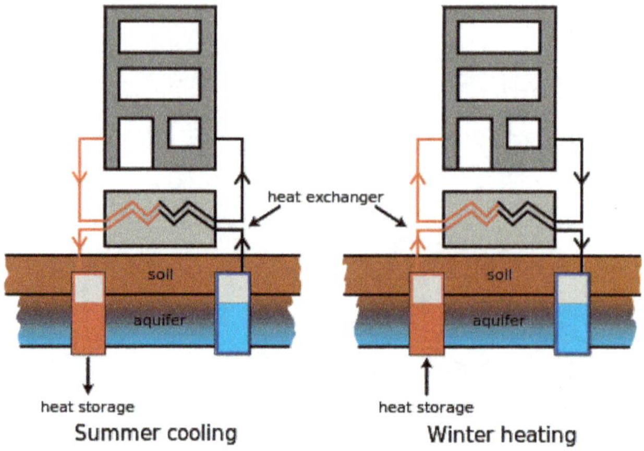

A heat pump in combination with heat and cold storage.

The efficiency of ground source heat pumps can be greatly improved by using seasonal thermal energy storage and interseasonal heat transfer. Heat captured and stored in thermal banks in the summer can be retrieved efficiently in the winter. Heat storage efficiency increases with scale, so this advantage is most significant in commercial or district heating systems.

Geosolar combisystems have been used to heat and cool a greenhouse using an aquifer for thermal storage. In summer, the greenhouse is cooled with cold ground water. This heats the water in the aquifer which can become a warm source for heating in winter. The combination of cold and heat storage with heat pumps can be combined with water/humidity regulation. These principles are used to provide renewable heat and renewable cooling to all kinds of buildings.

Also the efficiency of existing small heat pump installations can be improved by adding large, cheap, water filled solar collectors. These may be integrated into a to-be-overhauled parking lot, or in walls or roof constructions by installing one inch PE pipes into the outer layer.

Thermal Efficiency

The net thermal efficiency of a heat pump should take into account the efficiency of electricity generation and transmission, typically about 30%. Since a heat pump moves three to five times more heat energy than the electric energy it consumes, the total energy output is much greater than the electrical input. This results in net thermal efficiencies greater than 300% as compared to radiant electric heat being 100% efficient. Traditional combustion furnaces and electric heaters can never exceed 100% efficiency.

Geothermal heat pumps can reduce energy consumption— and corresponding air pollution emissions—up to 44% compared to air source heat pumps and up to 72% compared to electric resistance heating with standard air-conditioning equipment.

The dependence of net thermal efficiency on the electricity infrastructure tends to be an unnecessary complication for consumers and is not applicable to hydroelectric power, so performance of heat pumps is usually expressed as the ratio of heating output or heat removal to electricity input. Cooling performance is typically expressed in units of BTU/hr/watt as the energy efficiency ratio (EER), while heating performance is typically reduced to dimensionless units as the coefficient of performance (COP). The conversion factor is 3.41 BTU/hr/watt. Performance is influenced by all components of the installed system, including the soil conditions, the ground-coupled heat exchanger, the heat pump appliance, and the building distribution, but is largely determined by thelift" between the input temperature and the output temperature.

For the sake of comparing heat pump appliances to each other, independently from other system components, a few standard test conditions have been established by the American Refrigerant Institute (ARI) and more recently by the International Organization for Standardization. Standard ARI 330 ratings were intended for closed loop ground-source heat pumps, and assume secondary loop water temperatures of 77 °F (25 °C) for air conditioning and 32 °F (0 °C) for heating. These temperatures are typical of installations in the northern US. Standard ARI 325 ratings were intended for open loop ground-source heat pumps, and include two sets of ratings for groundwater temperatures of 50 °F (10 °C) and 70 °F (21 °C). ARI 325 budgets more electricity for water pumping than ARI 330. Neither of these standards attempt to account for seasonal variations. Standard ARI 870 ratings are intended for direct exchange ground-source heat pumps. ASHRAE transitioned to ISO 13256-1 in 2001, which replaces ARI 320, 325 and 330. The new ISO standard produces slightly higher ratings because it no longer budgets any electricity for water pumps.

Efficient compressors, variable speed compressors and larger heat exchangers all contribute to heat pump efficiency. Residential ground source heat pumps on the market today have standard COPs ranging from 2.4 to 5.0 and EERs ranging from 10.6 to 30. To qualify for an Energy Star label, heat pumps must meet certain minimum COP and EER ratings which depend on the ground heat exchanger type. For closed loop systems, the ISO 13256-1 heating COP must be 3.3 or greater and the cooling EER must be 14.1 or greater.

Actual installation conditions may produce better or worse efficiency than the standard test conditions. COP improves with a lower temperature difference between the input and output of the heat pump, so the stability of ground temperatures is important. If the loop field or water pump is undersized, the addition or removal of heat may push the ground temperature beyond standard test conditions, and performance will be degraded. Similarly, an undersized blower may allow the plenum coil to overheat and degrade performance.

Soil without artificial heat addition or subtraction and at depths of several metres or more remains at a relatively constant temperature year round. This temperature equates roughly to the average annual air-temperature of the chosen location, usually 7–12 °C

(45–54 °F) at a depth of 6 metres (20 ft) in the northern US. Because this temperature remains more constant than the air temperature throughout the seasons, geothermal heat pumps perform with far greater efficiency during extreme air temperatures than air conditioners and air-source heat pumps.

Standards ARI 210 and 240 define Seasonal Energy Efficiency Ratio (SEER) and Heating Seasonal Performance Factors (HSPF) to account for the impact of seasonal variations on air source heat pumps. These numbers are normally not applicable and should not be compared to ground source heat pump ratings. However, Natural Resources Canada has adapted this approach to calculate typical seasonally adjusted HSPFs for ground-source heat pumps in Canada. The NRC HSPFs ranged from 8.7 to 12.8 BTU/hr/watt (2.6 to 3.8 in nondimensional factors, or 255% to 375% seasonal average electricity utilization efficiency) for the most populated regions of Canada. When combined with the thermal efficiency of electricity, this corresponds to net average thermal efficiencies of 100% to 150%.

Environmental Impact

The US Environmental Protection Agency (EPA) has called ground source heat pumps the most energy-efficient, environmentally clean, and cost-effective space conditioning systems available. Heat pumps offer significant emission reductions potential, particularly where they are used for both heating and cooling and where the electricity is produced from renewable resources.

Ground-source heat pumps have unsurpassed thermal efficiencies and produce zero emissions locally, but their electricity supply includes components with high greenhouse gas emissions, unless the owner has opted for a 100% renewable energy supply. Their environmental impact therefore depends on the characteristics of the electricity supply and the available alternatives.

Annual greenhouse gas (GHG) savings from using a ground source heat pump instead of a high-efficiency furnace in a detached residence (assuming no specific supply of renewable energy)

Country	Electricity CO_2 Emissions Intensity	GHG savings relative to		
		natural gas	heating oil	electric heating
Canada	223 ton/GWh	2.7 ton/yr	5.3 ton/yr	3.4 ton/yr
Russia	351 ton/GWh	1.8 ton/yr	4.4 ton/yr	5.4 ton/yr
US	676 ton/GWh	-0.5 ton/yr	2.2 ton/yr	10.3 ton/yr
China	839 ton/GWh	-1.6 ton/yr	1.0 ton/yr	12.8 ton/yr

The GHG emissions savings from a heat pump over a conventional furnace can be calculated based on the following formula:

$$GHG\ Savings = HL\left(\frac{FI}{AFUE \times 1000\frac{kg}{ton}} - \frac{EI}{COP \times 3600\frac{sec}{hr}}\right)$$

- HL = seasonal heat load ≈ 80 GJ/yr for a modern detached house in the northern US

- FI = emissions intensity of fuel = 50 kg(CO_2)/GJ for natural gas, 73 for heating oil, 0 for 100% renewable energy such as wind, hydro, photovoltaic or solar thermal
- AFUE = furnace efficiency ≈ 95% for a modern condensing furnace
- COP = heat pump coefficient of performance ≈ 3.2 seasonally adjusted for northern US heat pump
- EI = emissions intensity of electricity ≈ 200-800 ton(CO_2)/GWh, depending on region

Ground-source heat pumps always produce fewer greenhouse gases than air conditioners, oil furnaces, and electric heating, but natural gas furnaces may be competitive depending on the greenhouse gas intensity of the local electricity supply. In countries like Canada and Russia with low emitting electricity infrastructure, a residential heat pump may save 5 tons of carbon dioxide per year relative to an oil furnace, or about as much as taking an average passenger car off the road. But in cities like Beijing or Pittsburgh that are highly reliant on coal for electricity production, a heat pump may result in 1 or 2 tons more carbon dioxide emissions than a natural gas furnace. For areas not served by utility natural gas infrastructure, however, no better alternative exists.

The fluids used in closed loops may be designed to be biodegradable and non-toxic, but the refrigerant used in the heat pump cabinet and in direct exchange loops was, until recently, chlorodifluoromethane, which is an ozone depleting substance. Although harmless while contained, leaks and improper end-of-life disposal contribute to enlarging the ozone hole. For new construction, this refrigerant is being phased out in favor of the ozone-friendly but potent greenhouse gas R410A. The EcoCute water heater is an air-source heat pump that uses carbon dioxide as its working fluid instead of chlorofluorocarbons. Open loop systems (i.e. those that draw ground water as opposed to closed loop systems using a borehole heat exchanger) need to be balanced by reinjecting the spent water. This prevents aquifer depletion and the contamination of soil or surface water with brine or other compounds from underground.

Before drilling, the underground geology needs to be understood, and drillers need to be prepared to seal the borehole, including preventing penetration of water between strata. The unfortunate example is a geothermal heating project in Staufen im Breisgau, Germany which seems the cause of considerable damage to historical buildings there. In 2008, the city centre was reported to have risen 12 cm, after initially sinking a few millimeters. The boring tapped a naturally pressurized aquifer, and via the borehole this water entered a layer of anhydrite, which expands when wet as it forms gypsum. The swelling will stop when the anhydrite is fully reacted, and reconstruction of the city centeris not expedient until the uplift ceases." By 2010 sealing of the borehole had not been accomplished. By 2010, some sections of town had risen by 30 cm.

Ground-source heat pump technology, like building orientation, is a natural building technique (bioclimatic building).

Economics

Ground source heat pumps are characterized by high capital costs and low operational costs compared to other HVAC systems. Their overall economic benefit depends primarily on the relative costs of electricity and fuels, which are highly variable over time and across the world. Based on recent prices, ground-source heat pumps currently have lower operational costs than any other conventional heating source almost everywhere in the world. Natural gas is the only fuel with competitive operational costs, and only in a handful of countries where it is exceptionally cheap, or where electricity is exceptionally expensive. In general, a homeowner may save anywhere from 20% to 60% annually on utilities by switching from an ordinary system to a ground-source system.

Capital costs and system lifespan have received much less study until recently, and the return on investment is highly variable. The most recent data from an analysis of 2011-2012 incentive payments in the state of Maryland showed an average cost of residential systems of $1.90 per watt, or about $26,700 for a typical (4 ton) home system. An older study found the total installed cost for a system with 10 kW (3 ton) thermal capacity for a detached rural residence in the US averaged $8000–$9000 in 1995 US dollars. More recent studies found an average cost of $14,000 in 2008 US dollars for the same size system. The US Department of Energy estimates a price of $7500 on its website, last updated in 2008. One source in Canada placed prices in the range of $30,000-$34,000 Canadian dollars. The rapid escalation in system price has been accompanied by rapid improvements in efficiency and reliability. Capital costs are known to benefit from economies of scale, particularly for open loop systems, so they are more cost-effective for larger commercial buildings and harsher climates. The initial cost can be two to five times that of a conventional heating system in most residential applications, new construction or existing. In retrofits, the cost of installation is affected by the size of living area, the home's age, insulation characteristics, the geology of the area, and location of the property. Proper duct system design and mechanical air exchange should be considered in the initial system cost.

Payback period for installing a ground source heat pump in a detached residence

Country	Payback period for replacing		
	natural gas	heating oil	electric heating
Canada	13 years	3 years	6 years
US	12 years	5 years	4 years
Germany	net loss	8 years	2 years

Notes:

- Highly variable with energy prices.
- Government subsidies not included.
- Climate differences not evaluated.

Capital costs may be offset by government subsidies; for example, Ontario offered $7000 for residential systems installed in the 2009 fiscal year. Some electric companies offer special rates to customers who install a ground-source heat pump for heating or cooling their building. Where electrical plants have larger loads during summer months and idle capacity in the winter, this increases electrical sales during the winter months. Heat pumps also lower the load peak during the summer due to the increased efficiency of heat pumps, thereby avoiding costly construction of new power plants. For the same reasons, other utility companies have started to pay for the installation of ground-source heat pumps at customer residences. They lease the systems to their customers for a monthly fee, at a net overall saving to the customer.

The lifespan of the system is longer than conventional heating and cooling systems. Good data on system lifespan is not yet available because the technology is too recent, but many early systems are still operational today after 25–30 years with routine maintenance. Most loop fields have warranties for 25 to 50 years and are expected to last at least 50 to 200 years. Ground-source heat pumps use electricity for heating the house. The higher investment above conventional oil, propane or electric systems may be returned in energy savings in 2–10 years for residential systems in the US. If compared to natural gas systems, the payback period can be much longer or non-existent. The payback period for larger commercial systems in the US is 1–5 years, even when compared to natural gas. Additionally, because geothermal heat pumps usually have no outdoor compressors or cooling towers, the risk of vandalism is reduced or eliminated, potentially extending a system's lifespan.

Ground source heat pumps are recognized as one of the most efficient heating and cooling systems on the market. They are often the second-most cost effective solution in extreme climates (after co-generation), despite reductions in thermal efficiency due to ground temperature. (The ground source is warmer in climates that need strong air conditioning, and cooler in climates that need strong heating.)

Commercial systems maintenance costs in the US have historically been between \$0.11 to \$0.22 per m^2 per year in 1996 dollars, much less than the average \$0.54 per m^2 per year for conventional HVAC systems.

Governments that promote renewable energy will likely offer incentives for the consumer (residential), or industrial markets. For example, in the United States, incentives are offered both on the state and federal levels of government. In the United Kingdom the Renewable Heat Incentive provides a financial incentive for generation of renewable heat based on metered readings on an annual basis for 20 years for commercial buildings. The domestic Renewable Heat Incentive is due to be introduced in Spring 2014 for seven years and be based on deemed heat.

Installation

Because of the technical knowledge and equipment needed to design and size the system properly (and install the piping if heat fusion is required), a GSHP system installation requires a professional's services. Several installers have published real-time views of system performance in an online community of recent residential installations. The International Ground Source Heat Pump Association (IGSHPA), Geothermal Exchange Organization (GEO), the Canadian GeoExchange Coalition and the Ground Source Heat Pump Association maintain listings of qualified installers in the US, Canada and the UK. Furthermore, detailed analysis of Soil thermal conductivity for horizontal systems and formation thermal conductivity for vertical systems will generally result in more accurately design systems with a higher efficiency.

REFERENCES

1. National Non-Food Crops Centre. "NNFCC Renewable Fuels and Energy Factsheet: Anaerobic Digestion", Retrieved on 2011-11-22
2. Koyama, Tadashiro (1963).Gaseous metabolism in lake sediments and paddy soils and the production of atmospheric methane and hydrogen". Journal of Geophysical Research. 68 (13): 3971–3973. doi:10.1029/JZ068i013p03971.
3. Pamatmat, Mario Macalalag; Bhagwat, Ashok M. (1973).Anaerobic metabolism in Lake Washington sediments" (PDF). Limnology and Oceanography. pp. 611–627. doi:10.4319/lo.1973.18.4.0611. Archived (PDF) from the original on 16 December 2013.
4. Zehnder, Alexander J. B. (1978).Ecology of methane formation". In Mitchell, Ralph. Water pollution microbiology 2. New York: Wiley. pp. 349–376. ISBN 978-0-471-01902-2.
5. MacGregor, A. N.; Keeney, D. R. (1973).Methane formation by lake sediments during in vitro incubations". Journal of the American Water Resources Association (JAWRA). 9 (6): 1153–1158. doi:10.1111/j.1752-1688.1973.tb05854.x.
6. Anaerobic digestion reference sheet, waste.nl. Retrieved 25.10.07.
7. Meisam Tabatabaei, Raha Abdul Rahim, André-Denis G. Wright, Yoshihito Shirai, Norhani Abdullah, Alawi Sulaiman, Kenji Sakai and Mohd Ali Hassan. 2010. Importance of the methanogenic archaea populations in anaerobic wastewater treatments (Process Biochemistry- 45(8), pp: 1214-1225)
8. Agricultural Biogas, www.clarke-energy.com, accessed 08.11.11
9. GE Jenbacher Biogas Engines, www.clarke-energy.com, accessed 15.04.11
10. Anaerobic Digestion Strategy and Action Plan, defra.gov.uk. Accessed 19.01.2012
11. (source needed)
12. Jyllands-Posten 29. December 2011. Accessed 19.01.2012 via Google Translate
13. National Non-Food Crops Centre. Evaluation of Opportunities for Converting Indigenous UK Wastes to Fuels and Energy (Report), NNFCC 09-012
14. Adapted from Beychok, M. (1967) Aqueous Wastes from Petroleum and Petrochemical Plants, First edition, John Wiley & Sons, LCCN 67019834
15. Fergusen, T. & Mah, R. (2006) Methanogenic bacteria in Anaerobic digestion of biomass, p49
16. The biogas plant, unu.edu. Retrieved 5.11.07.
17. Anaerobic digestion, waste.nl. Retrieved 19.08.07.
18. Sleat, R. & Mah, R. (2006) Hydrolytic Bacteria in Anaerobic digestion of biomass, p15
19. Boone, D. & Mah, R. (2006) Transitional bacteria in anaerobic digestion of

biomass, p35

20. What is anaerobic digestion, sop.inria.fr. Retrieved 24.10.07.
21. Anaerobic digestion, biotank.co.uk. Retrieved 24.10.07.
22. Anaerobic Digester: How It Works [Process] | GreenTheFuture.com". greenthefuture.com. Retrieved 2016-02-23.
23. Landfill Gas & Biogas Analyzers | Nova Gas". Nova Gas. Retrieved 2016-02-23.
24. A, Igoni Hilkia; Abowei, M. F. N.; Ayotamuno, M. J.; Eze, C. L. (2009-01-16).Comparative Evaluation of Batch and Continuous Anaerobic Digesters in Biogas Production from Municipal Solid Waste using Mathematical Models". Agricultural Engineering International: CIGR Journal. ISSN 1682-1130.
25. Comparing Of Mesophilic And Thermophilic Anaerobic Fermented Sewage Sludge Based On Chemical And Biochemical Tests" . www.aloki.hu. Retrieved 23 February 2016.
26. Low and High Solid Anaerobic Digestion Technology". www.theecoambassador.com. Retrieved 2016-02-23.
27. 2008, DEVinitiv GbR,. "Anaerobic Digestion Systems". www.wtert.eu. Retrieved 2016-02-23.
28. aikantechnology.com Retrieved 10. Feb. 2012.
29. Anaerobic digestion, energy.ca.gov. Retrieved 18.06.09.
30. BIOPAQ IC, paques.nl. Retrieved 19.08.07.
31. Biological processes with Biomar technology envirochemie.com. Retrieved 24.10.2012.
32. Song, Y.C., Kwon, S.J., Woo, J.H. (2004) Mesophilic and thermophilic temperature co-phase anaerobic digestion compared with single-stage mesophilic- and thermophilic digestion of sewage sludge, Water Res. 2004 Apr;38(7):1653–62
33. Transfer of low-cost plastic biodigester technology at household level in Bolivia, lrrd.org
34. Gupta, Sujata (2010-11-06).Biogas comes in from the cold". New Scientist. London: Sunita Harrington. p. 14. Retrieved 2011-02-04.
35. Animal by-products introduction, ec.europa.eu. Retrieved 24.10.07.
36. Feasibility study concerning anaerobic digestion in Northern Ireland, eunomia.co.uk , Retrieved 19.08.07. Archived 28 November 2007 at the Wayback Machine.
37. Jewell, W.; Cummings, R.; Richards, B. (1993). "Methane fermentation of energy crops: Maximum conversion kinetics and in situ biogas purification". Biomass and Bioenergy. 5 (3–4): 261–278. doi:10.1016/0961-9534(93)90076-G.
38. Richards, B.; Cummings, R.; White, T.; Jewell, W. (1991).Methods for kinetic analysis of methane fermentation in high solids biomass digesters". Biomass and Bioenergy. 1 (2): 65–26. doi:10.1016/0961-9534(91)90028-B.
39. Biomethanation in advances in biochemical engineering and biotechnology,

books.google.com. Retrieved 24.10.07.

40. Anaerobic Lagoons for Storage/Treatment of Livestock Manure, missouri.edu. Retrieved 8.11.07.
41. Abstract: Methanogenic population dynamics during start-up of anaerobic digesters treating municipal solid waste and biosolids, interscience.wiley.com. Retrieved 24.10.07.
42. Animal By-Products Regulations, defra.gov.uk. Retrieved 24.10.07. Archived16 April 2014 at the Wayback Machine.
43. HIMET—A Two-Stage Anaerobic Digestion Process for Converting Waste to Energy, gastechnology.org. Retrieved 19.08.07.
44. Finstein, M. S. (2006) ArrowBio process integrates preprocessing and advanced anaerobic digestion to recover recyclables and generate electricity, oaktech-environmental.com. Retrieved 19.08.07.
45. Chen Y., Cheng J.J., Creamer K.S. (2008). Inhibition of anaerobic digestion process: A review. Bioresource Technology 99, 4044-4064
46. Alfagy.com, retrieved 16.08.09
47. Anaerobic digestion feedstock classification, wisbiorefine.org. Retrieved 24.10.07.
48. Lemmer, A. & Oeschsner, H. Co-fermentation of grass and forage maize, Energy, Landtechnik, 5/11, p 56, ltnet.lv-h.de
49. , waste-management-world.com. Retrieved 24.10.07.
50. Benner, Ronald (1989).Book Review: Biology of anaerobic microorganisms"(PDF). Limnology and Oceanography. 34 (3): 647. doi:10.4319/lo.1989.34.3.0647. Archived from the original (PDF) on 13 November 2006.
51. Anaerobic Digestion Initiative Advisory Committee (ADIAC).Feedstock". Archived from the original on 13 December 2011.
52. Richards, B.; Cummings, R. J.; Jewell, W. J. (1991).High rate low solids methane fermentation of sorghum, corn and cellulose". Biomass and Bioenergy. 1 (5): 249–260. doi:10.1016/0961-9534(91)90036-C.
53. Richards, B.; Cummings, R. J.; Jewell, W. J.; Herndon, F. G. (1991).High solids anaerobic methane fermentation of sorghum and cellulose". Biomass and Bioenergy. 1: 47–53. doi:10.1016/0961-9534(91)90051-D.
54. National Non-Food Crops Centre. Farm-Scale Anaerobic Digestion Plant Efficiency, NNFCC 11-015
55. Management of Urban Biodegradable Waste, books.google.com. Retrieved 24.10.07.
56. Anaerobic co-digestion of sewage sludge and rice straw, bvsde.ops-oms.org. Retrieved 24.10.07.
57. Anaerobic digestion of classified municipal solid wastes, seas.ucla.edu. Retrieved 24.10.07.
58. National Non-Food Crops Centre. Economic Assessment of Anaerobic Digestion Technology & its Suitability to UK Farming & Waste Systems (Report, 2nd

Edition), NNFCC 10-010

59. Jerger, D. & Tsao, G. (2006) Feed composition in Anaerobic digestion of biomass, p65

60. Tilley, E., Ulrich, L., Lüthi, C., Reymond, Ph., Zurbrügg, C. (2014) Compendium of Sanitation Systems and Technologies - (2nd Revised Edition). Swiss Federal Institute of Aquatic Science and Technology (Eawag), Duebendorf, Switzerland.

61. Anaerobic Digestion, wasteresearch.co.uk. Retrieved 24.10.07.

62. Sea Dumping of Sewage Sludge, encyclopedia.com. Retrieved 22.02.2010.

63. Ocean Dumping Ban Act (1988), bookrags.com. Retrieved 22.02.2010.

64. Juniper (2005) MBT: A Guide for Decision Makers – Processes, Policies & Markets, juniper.co.uk, (Project funding supplied by Sita Environmental Trust). Retrieved 22.11.06.

65. Svoboda, I (2003) Anaerobic digestion, storage, olygolysis, lime, heat and aerobic treatment of livestock manures, scotland.gov.uk. Retrieved 17.08.07.

66. Haase Mechanical Biological Treatment and Wet Anaerobic Digestion, haase-energietechnik.de. Retrieved 23.10.07.

67. Global warming methane could be far more potent than carbon dioxidenewmediaexplorer.org. Retrieved 17.08.07.

68. Renewable Energy Framework, esru.strath.ac.uk. Retrieved 8.11.07.

69. Friends of the Earth (2004) Anaerobic digestion Briefing Paper, foe.co.uk. Retrieved 17.08.07.

70. Cardiff University (2005) Anaerobic Digestion Page, wasteresearch.co.uk. Retrieved 17.08.07.

71. Doelle, H. W. (2001) Biotechnology and Human Development in Developing Countries, ejbiotechnology.info. Retrieved 19.08.07.

72. Biogas Bonanza for Third World Development, i-sis.org.uk. Retrieved 4.11.07.

73. The Clean Development Mechanism in Nepal in The Tiempo Climate Newswatch, tiempocyberclimate.org

74. Benefits of Anaerobic Digestion, afbini.gov.uk. Retrieved 22 February 2010. Archived 9 May 2013 at the Wayback Machine.

75. Questions about biomass energy, dti.gov.uk. Retrieved 17.08.07.

76. 38% HHV Caterpillar Bio-gas Engine Fitted to Sewage Works | Claverton Group, claverton-energy.com

77. Alfagy.com, "Be Green – Make Gas"

78. CHP Feed In Tariffs & Green Energy Financial Support, www.alfagy.com

79. East Bay Municipal Utility District (2008). Anaerobic Digestion of Food Waste(PDF). United States Environmental Protection Agency.

80. Organics: Anaerobic Digestion". United States Environmental Protection Agency. Retrieved 16 September 2013.

81. Half Britain's homes could be heated by renewable gas, nationalgrid.com

82. Petersson A., Wellinger A. (2009). Biogas upgrading technologies - developments and innovations. IEA Bioenergy Task 37

83. Biogas flows through Germany's grid 'big time', renewableenergyworld.com

84. Transmission loss, energyvortex.com

85. Shah, Dhruti (5 October 2010).Oxfordshire town sees human waste used to heat homes". BBC NEWS. Archived from the original on 5 October 2010. Retrieved 5 October 2010.

86. https://www.theguardian.com/environment/2015/apr/20/ecotricity-offers-anaerobic-digestion-of-grass-as-answer-to-fracking

87. http://www.hydrogenfuelnews.com/green-energy-plant-could-be-uks-alternative-to-fracking/8521901/

88. Natural & bioGas Vehicle Association (NVGA). Sweden

89. Introduction and Spanish organic waste situation, compostnetwork.info. Retrieved 19.08.07.

90. Satisfaction of reducing your carbon footprint". The Hindu. Retrieved 2012-07-31.

91. Abstract from Operation of Municipal Wastewater Treatment Plants Manual of Practice-MOP 11 Fifth Edition, e-wef.org. Retrieved 19.08.07.

92. Anaerobic Digestion – An Introduction and Commercial Status in the US – As of 2006, anaerobic-digestion.com. Retrieved 07.12.14

93. Basic Information on Biogas, www.kolumbus.fi. Retrieved 2.11.07.

94. Roubík, Hynek; Mazancová, Jana; Banout, Jan; Verner, Vladimír (2016-01-20). Addressing problems at small-scale biogas plants: a case study from central Vietnam". Journal of Cleaner Production. 112, Part 4: 2784–2792. doi:10.1016/j.jclepro.2015.09.114.

95. Beginners Guide to Biogas. Retrieved 04.10.15.

96. How Anaerobic Digestion (Methane Recovery) Works, eere.energy.gov. Retrieved 19.08.07.

97. Anaerobic digestion briefing sheet, foe.co.uk. Retrieved 24.10.07.

98. GE Energy – Jenbacher Gas Engines for Power Generation, power-technology.com. Retrieved 19 August 2007.

99. work3.pdf UK Biomass Strategy 2007, defra.gov.uk , Retrieved 19.08.07. Archived 16 December 2008 at the Wayback Machine.

100. What is anaerobic digestion?, afbini.gov.uk. Retrieved 24.10.07.

101. Removal of hydrogen sulfide from anaerobic digester gas, U.S. Patent, patentstorm.us. Retrieved 17.08.07.

102. Abstract from Online Measurement of Dissolved and Gaseous-Hydrogen Sulfide in Anaerobic Biogas Reactors, cheric.org. Retrieved 24.10.07.

103. Wheles, E. & Pierece, E. (2004) Siloxanes in landfill and digester gas, scsengineers.com. Retrieved 17.08.07.

104. Biogas Upgrading and Utilisation, EEA Bioenergy, iea-biogas.net. Retrieved 25.10.07.

105. Tower, P.; Wetzel, J.; Lombard, X. (March 2006).New Landfill Gas Treatment Technology Dramatically Lowers Energy Production Costs". Applied Filter Technology. Archived from the original on 24 September 2011. Retrieved 2009-04-30., appliedfiltertechnology.com

106. Richards, B.; Herndon, F. G.; Jewell, W. J.; Cummings, R. J.; White, T. E. (1994).In situ methane enrichment in methanogenic energy crop digesters". Biomass and Bioenergy. 6 (4): 275–274. doi:10.1016/0961-9534(94)90067-1.

107. Biogas as a road transport fuel nfuonline.com. Retrieved 24.10.07. Archived15 October 2007 at the Wayback Machine.

108. Haase biogas energy centre haase-energietechnik.de. Retrieved 19.08.07. Archived 17 December 2008 at the Wayback Machine.

109. Fact sheet on anaerobic digestion, waste.nl. Retrieved 19.08.07.

110. Biomass and biogas globalwarming101.com. Retrieved 19.08.07. Archived21 August 2007 at the Wayback Machine.

111. Oaktech Consultation Response to UK Source Segregation Requirement, alexmarshall.me.uk. Retrieved 19.08.07.

112. UK Strategy for centralised anaerobic digestion, ingentaconnect.com. Retrieved 24.10.07.

113. Solid Digestate to Ethanol, onlinelibrary.wiley.com Retrieved 11.18.10

114. Biomass and biogas, globalwarming101.com. Retrieved 24.10.07. Archived 5 June 2007 at the Wayback Machine.

115. Vitoria Plant Information, ows.be. Retrieved 24.10.07.

116. Kompogas Homepage, kompogas.ch. Retrieved 24.10.07. Archived 9 February 2008 at the Wayback Machine.

117. Dosta, Joan; Galí, Alexandre; Macé, Sandra; Mata-Álvarez, Joan (February 2007). Modelling a sequencing batch reactor to treat the supernatant from anaerobic digestion of the organic fraction of municipal solid waste". Journal of Chemical Technology & Biotechnology. 82 (2): 158–64. doi:10.1002/jctb.1645. Retrieved 16 September 2013.

118. Clarke Energy Reverse Osmosis Unit, clarke-energy.co.uk. Retrieved 24.10.07. Archived 16 December 2007 at the Wayback Machine.

119. BOD Effluent Treatment, virtualviz.com. Retrieved 24.10.07.

120. Cruazon, B. (2007) History of anaerobic digestion, web.pdx.edu. Retrieved 17.08.07.

121. Water.me.vccs.edu. Retrieved 22 February 2010.

122. Humenik, F. *et al.* (2007) , epa.gov. Retrieved 14.07.14.

CHAPTER 7

Bioconversion Systems

Chapter Contents

Bioconversion, also known as biotransformation refers to the use of live organisms often microorganisms to carry out a chemical reaction that is more costly or not feasible no biologically. These organisms convert a substance to a chemically modified form. Microbial conversion of petroleum resins existing in a heavy fraction of crude oil was investigated by native isolated consortia. The heavy oil fraction was taken from an end cut of distillation tower of Abadan refinery in Iran. Several Soil samples were taken from different oil contaminated sites of Abadan oil refinery in Khuzestan province. Seven bacterial consortia were then prepared by enrichment method and were examined for their ability to convert resins.

INTRODUCTION

Bioconversion, also known as *biotransformation*, is the conversion of organic materials, such as plant or animal waste, into usable products or energy sources by biological processes or agents, such as certain microorganisms. One example is the industrial production of cortisone, which one step is the bioconversion of progesterone to 11-alpha-Hydroxyprogesterone by

Rhizopus nigricans. Another example is the bioconversion of glycerol to 1,3-propanediol, which is part of scientific research for many decades.

Another example of bioconversion is the conversion of organic materials, such as plant or animal waste, into usable products or energy sources by biological processes or agents, such as certain microorganisms, some detritivores or enzymes.

In the USA, the Bioconversion Science and Technology group performs multidisciplinary R&D for the Department of Energy's (DOE) relevant applications of bioprocessing, especially with biomass. Bioprocessing combines the disciplines of chemical engineering, microbiology and biochemistry. The Group ‹s primary role is investigation of the use of microorganism, microbial consortia and microbial enzymes in bioenergy research. New cellulosic ethanol conversion processes have enabled the variety and volume of feedstock that can be bioconverted to expand rapidly. Feedstock now includes materials derived from plant or animal waste such as paper, auto-fluff, tires, fabric, construction materials, municipal solid waste (MSW), sludge, sewage, etc.

Three different processes for bioconversion

1 - **Enzymatic hydrolysis** - a single source of feedstock, switchgrass for example, is mixed with strong enzymes which convert a portion of cellulosic material into sugars which can then be fermented into ethanol. Genencor and Novozymes are two companies that have received United States government Department of Energy funding for research into reducing the cost of cellulase, a key enzyme in the production cellulosic ethanol by this process.

2 - **Synthesis gas fermentation** - a blend of feedstock, not exceeding 30% water, is gasified in a closed environment into a syngas containing mostly carbon monoxide and hydrogen. The cooled syngas is then converted into usable products through exposure to bacteria or other catalysts. BRI Energy, LLC is a company whose pilot plant in Fayetteville, Arkansas is currently using synthesis gas fermentation to convert a variety of waste into ethanol. After gasification, anaerobic bacteria (*Clostridium ljungdahlii*) are used to convert the syngas (CO, CO_2, and H_2) into ethanol. The heat generated by gasification is also used to co-generate excess electricity.

3 - **C.O.R.S. and Grub Composting** are sustainable technologies that employ organisms that feed on organic matter to reduce and convert organic waste in to high quality feedstuff and oil rich material for the biodiesel industry. Organizations pioneering this novel approach to waste management are EAWAG, ESR International, Prota Culture and BIOCONVERSION that created the *e*-CORS® system to meet large scale organic waste management needs and environmental sustainability in both urban and livestock farming reality. This type of engineered system introduces a substantial innovation represented by the automatic modulation of the treatment, able to adapt conditions of the system to the biology of the scavenger used, improving their performances and the power of this technology.

BIODRYING

Biodrying is the process by which biodegradable waste is rapidly heated through initial stages of composting to remove moisture from a waste stream and hence reduce its overall weight. In biodrying processes, the drying rates are augmented by biological heat in addition to forced aeration. The major portion of biological heat, naturally available through the aerobic degradation of organic matter, is utilized to evaporate surface and bound water associated with the mixed sludge. This heat generation assists in reducing the moisture content of the biomass without the need for supplementary fossil fuels, and with minimal electricity consumption. It can take as little as 8 days to dry waste in this manner. This enables reduced costs of disposal if landfill is charged on a cost per tonne basis. Biodrying may be used as part of the production process for refuse-derived fuels. Biodrying does not however greatly affect the biodegradability of the waste and hence is not stabilised. Biodried waste will still break down in a landfill to produce landfill gas and hence potentially contribute to climate change. In the UK this waste will still impact upon councils LATS allowances. Whilst biodrying is increasingly applied within commercial mechanical biological treatment (MBT) plants, it is also still subject to on-going research and development.

BIOCONVERSION OF BIOMASS TO MIXED ALCOHOL FUELS

The **bioconversion of biomass to mixed alcohol fuels** can be accomplished using the MixAlco process. Through bioconversion of biomass to a mixed alcohol fuel, more energy from the biomass will end up as liquid fuels than in converting biomass to ethanol by yeast fermentation.

The process involves a biological/chemical method for converting any biodegradable material (e.g., urban wastes, such as municipal solid waste, biodegradable waste, and sewage sludge, agricultural residues such as corn stover, sugarcane bagasse, cotton gin trash, manure) into useful chemicals, such as carboxylic acids (e.g., acetic, propionic, butyric acid), ketones (e.g., acetone, methyl ethyl ketone, diethyl ketone) and biofuels, such as a mixture of primary alcohols (e.g., ethanol, propanol, *n*-butanol) and/or a mixture of secondary alcohols (e.g., isopropanol, 2-butanol, 3-pentanol). Because of the many products that can be economically produced, this process is a true biorefinery.

The process uses a mixed culture of naturally occurring microorganisms found in natural habitats such as the rumen of cattle, termite guts, and marine and terrestrial swamps to anaerobically digest biomass into a mixture of carboxylic acids produced during the acidogenic and acetogenic stages of anaerobic digestion, however with the inhibition of the methanogenic final stage. The more popular methods for production of ethanol and cellulosic ethanol use enzymes that must be isolated first to be added to the biomass and thus convert the starch or cellulose into simple sugars, followed then by yeast fermentation into ethanol. This process does not need the addition of such enzymes as these microorganisms make their own.

Pilot Plant (College Station, Texas).

As the microoganisms anaerobically digest the biomass and convert it into a mixture of carboxylic acids, the pH must be controlled. This is done by the addition of a buffering agent (e.g., ammonium bicarbonate, calcium carbonate), thus yielding a mixture of carboxylate salts. Methanogenesis, being the natural final stage of anaerobic digestion, is inhibited by the presence of the ammonium ions or by the addition of an inhibitor (e.g., iodoform). The resulting fermentation broth contains the produced carboxylate salts that must be dewatered. This is achieved efficiently by vapor-compression evaporation. Further chemical refining of the dewatered fermentation broth may then take place depending on the final chemical or biofuel product desired.

The condensed distilled water from the vapor-compression evaporation system is recycled back to the fermentation. On the other hand, if raw sewage or other waste water with high BOD in need of treatment is used as the water for the fermentation, the condensed distilled water from the evaporation can be recycled back to the city or to the original source of the high-BOD waste water. Thus, this process can also serve as a water treatment facility, while producing valuable chemicals or biofuels.

Because the system uses a mixed culture of microorganisms, besides not needing any enzyme addition, the fermentation requires no sterility or aseptic conditions, making this front step in the process more economical than in more popular methods for the production of cellulosic ethanol. These savings in the front end of the process, where volumes are large, allows flexibility for further chemical transformations after dewatering, where volumes are small.

Carboxylic acids

Carboxylic acids can be regenerated from the carboxylate salts using a process known as "acid springing". This process makes use of a high-molecular-weight tertiary amine (e.g., trioctylamine), which is switched with the cation (e.g., ammonium or calcium). The resulting amine carboxylate can then be thermally decomposed into the amine itself, which is recycled, and the corresponding carboxylic acid. In this way, theoretically, no chemicals are consumed or wastes produced during this step.

Ketones

There are two methods for making ketones. The first one consists on thermally converting calcium carboxylate salts into the corresponding ketones. This was a common method for making acetone from calcium acetate during World War I. The other method for making ketones consists on converting the vaporized carboxylic acids on a catalytic bed of zirconium oxide.

Alcohols

Primary alcohols

The undigested residue from the fermentation may be used in gasification to make hydrogen (H_2). This H_2 can then be used to hydrogenolyze the esters over a catalyst (e.g., copper chromite), which are produced by esterifying either the ammonium carboxylate salts (e.g., ammonium acetate, propionate, butyrate) or the carboxylic acids (e.g., acetic, propionic, butyric acid) with a high-molecular-weight alcohol (e.g., hexanol, heptanol). From the hydrogenolysis, the final products are the high-molecular-weight alcohol, which is recycled back to the esterification, and the corresponding primary alcohols (e.g., ethanol, propanol, butanol).

Secondary Alcohols

The secondary alcohols (e.g., isopropanol, 2-butanol, 3-pentanol) are obtained by hydrogenating over a catalyst (e.g., Raney nickel) the corresponding ketones (e.g., acetone, methyl ethyl ketone, diethyl ketone).

Drop-in Biofuels

The primary or secondary alcohols obtained as described above may undergo conversion to drop-in biofuels, fuels which are compatible with current fossil fuel infrastructure such as biogasoline, green diesel and bio-jet fuel. Such is done by subjecting the alcohols to dehydration followed by oligomerization using zeolite catalysts in a manner similar to the methanex process, which used to produce gasoline from methanol in New Zealand.

Acetic acid Versus Ethanol

Cellulosic-ethanol manufacturing plants are bound to be net exporters of electricity because a large portion of the lignocellulosic biomass, namely lignin, remains undigested and it must be burned, thus producing electricity for the plant and excess electricity for the grid. As the market grows and this technology becomes more widespread, coupling the liquid fuel and the electricity markets will become more and more difficult.

Acetic acid, unlike ethanol, is biologically produced from simple sugars without the production of carbon dioxide:

$C_6H_{12}O_6 \rightarrow 2\ CH_3CH_2OH + 2\ CO_2$

(Biological production of ethanol)

$C_6H_{12}O_6 \rightarrow 3\ CH_3COOH$

(Biological production of acetic acid)

Because of this, on a mass basis, the yields will be higher than in ethanol fermentation. If then, the undigested residue (mostly lignin) is used to produce hydrogen by gasification, it is ensured that more energy from the biomass will end up as liquid fuels rather than excess heat/electricity.

$3\ CH_3COOH + 6\ H_2 \rightarrow 3\ CH_3CH_2OH + 3\ H_2O$

(Hydrogenation of acetic acid)

$C_6H_{12}O_6$ (from cellulose) + $6\ H_2$ (from lignin) $\rightarrow 3\ CH_3CH_2OH + 3\ H_2O$

(Overall reaction)

A more comprehensive description of the economics of each of the fuels is given on the pages alcohol fuel and ethanol fuel, more information about the economics of various systems can be found on the central page biofuel.

Stage of Development

The system has been in development since 1991, moving from the laboratory scale (10 g/day) to the pilot scale (200 lb/day) in 2001. A small demonstration-scale plant (5 ton/day) has been constructed and is under operation and a 220 ton/day demonstration plant is expected in 2012.

Waste Management

Waste management in Kathmandu, Nepal.

Waste management in Stockholm, Sweden.

Waste management or **Waste disposal** is all the activities and actions required to manage waste from its inception to its final disposal. This includes amongst other things, collection, transport, treatment and disposal of waste together with monitoring and regulation. It also encompasses the legal and regulatory framework that relates to waste management encompassing guidance on recycling etc.

The term normally relates to all kinds of waste, whether generated during the extraction of raw materials, the processing of raw materials into intermediate and final products, the consumption of final products, or other human activities, including municipal (residential, institutional, commercial), agricultural, and social (health care, household hazardous waste, sewage sludge). Waste management is intended to reduce adverse effects of waste on health, the environment or aesthetics.

Waste management practices are not uniform among countries (developed and developing nations); regions (urban and rural area), and sectors (residential and industrial).

Central Principles of Waste Management

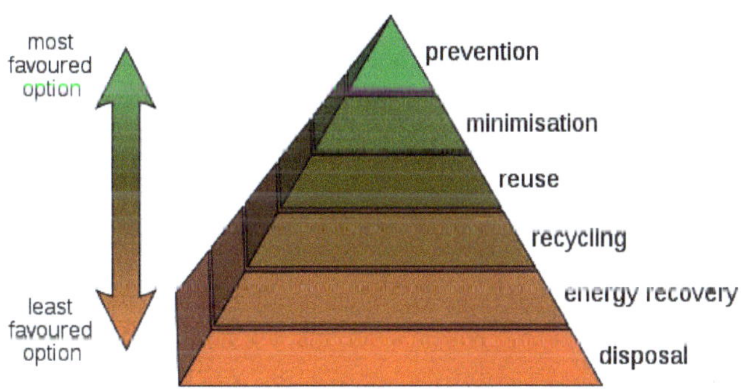

Diagram of the waste hierarchy.

There are a number of concepts about waste management which vary in their usage between countries or regions. Some of the most general, widely used concepts include:

Waste Hierarchy

The waste hierarchy refers to the "3 Rs reduce, reuse and recycle, which classify waste management strategies according to their desirability in terms of waste minimisation. The waste hierarchy remains the cornerstone of most waste minimisation strategies. The aim of the waste hierarchy is to extract the maximum practical benefits from products and to generate the minimum amount of waste; see: resource recovery.The waste hierarchy is represented as a pyramid because the basic premise is for policy to take action first and prevent the generation of waste. The next step or preferred action is to reduce the generation of waste i.e. by re-use. The next is recycling which would include composting. Following this step is material recovery and waste-to-energy. Energy can be recovered from processes i.e. landfill and combustion, at this level of the hierarchy. The final action is disposal, in landfills or through incineration without energy recovery. This last step is the final resort for waste which has not been prevented, diverted or recovered. The waste hierarchy represents the progression of a product or material through the sequential stages of the pyramid of waste management. The hierarchy represents the latter parts of the life-cycle for each product.

Life-cycle of a product

The life-cycle begins with design, then proceeds through manufacture, distribution, use and then follows through the waste hierarchy's stages of reduce, reuse and recycle. Each of the above stages of the life-cycle offers opportunities for policy intervention, to rethink the need for the product, to redesign to minimize waste potential, to extend its use. The key behind the life-cycle of a product is to optimize the use of the world's limited resources by avoiding the unnecessary generation of waste.

Resource efficiency

Resource efficiency reflects the understanding that current, global, economic growth and development can not be sustained with the current production and consumption patterns. Globally, we are extracting more resources to produce goods than the planet can replenish. Resource efficiency is the reduction of the environmental impact from the production and consumption of these goods, from final raw material extraction to last use and disposal. This process of resource efficiency can address sustainability.

Polluter Pays Principle

The Polluter pays principle is a principle where the polluting party pays for the impact caused to the environment. With respect to waste management, this generally refers to the requirement for a waste generator to pay for appropriate disposal of the unrecoverable material.

Throughout most of history, the amount of waste generated by humans was insignificant due to low population density and low societal levels of the exploitation of natural resources. Common waste produced during pre-modern times was mainly ashes and human biodegradable waste, and these were released back into the ground locally, with minimum environmental impact. Tools made out of wood or metal were generally reused or passed down through the generations.

However, some civilizations do seem to have been more profligate in their waste output than others. In particular, the Maya of Central America had a fixed monthly ritual, in which the people of the village would gather together and burn their rubbish in large dumps.

Modern Era

Sir Edwin Chadwick's 1842 report *The Sanitary Condition of the Labouring Population* was influential in securing the passage of the first legislation aimed at waste clearance and disposal.

Following the onset of industrialisation and the sustained urban growth of large population centres in England, the buildup of waste in the cities caused a rapid deterioration in levels of sanitation and the general quality of urban life. The streets became choked with filth due to the lack of waste clearance regulations. Calls for the establishment of a municipal authority with waste removal powers occurred as early as 1751, when Corbyn Morris in London proposed that "... as the preservation of the health of the people is of great importance, it is proposed that the cleaning of this city, should be put under one uniform public management, and all the filth be...conveyed by the Thames to proper distance in the country".

However, it was not until the mid-19th century, spurred by increasingly devastating cholera outbreaks and the emergence of a public health debate that the first legislation on the issue emerged. Highly influential in this new focus was the report *The Sanitary Condition of the Labouring Population* in 1842 of the social reformer, Edwin Chadwick, in which he argued for the importance of adequate waste removal and management facilities to improve the health and wellbeing of the city's population.

In the UK, the Nuisance Removal and Disease Prevention Act of 1846 began what was to be a steadily evolving process of the provision of regulated waste management in London. The Metropolitan Board of Works was the first citywide authority that centralized sanitation regulation for the rapidly expanding city and the Public Health Act 1875 made it compulsory for every household to deposit their weekly waste in "moveable receptacles: for disposal—the first concept for a dust-bin.

Manlove, Alliott & Co. Ltd. 1894 destructor furnace. The use of incinerators for waste disposal became popular in the late 19th century.

The dramatic increase in waste for disposal led to the creation of the first incineration plants, or, as they were then called, "destructors". In 1874, the first incinerator was built in Nottingham by Manlove, Alliott & Co. Ltd. to the design of Albert Fryer. However, these were met with opposition on account of the large amounts of ash they produced and which wafted over the neighbouring areas.

Similar municipal systems of waste disposal sprung up at the turn of the 20th century in other large cities of Europe and North America. In 1895, New York City became the first U.S. city with public-sector garbage management.

Early garbage removal trucks were simply open bodied dump trucks pulled by a team of horses. They became motorized in the early part of the 20th century and the first close body trucks to eliminate odours with a dumping lever mechanism were introduced in the 1920s in Britain. These were soon equipped with ‹hopper mechanisms' where the scooper was loaded at floor level and then hoisted mechanically to deposit the waste in the truck. The Garwood Load Packer was the first truck in 1938, to incorporate a hydraulic compactor.

Waste Handling and Transport

Molded plastic, wheeled waste bin in Berkshire, England.

Waste collection methods vary widely among different countries and regions. Domestic waste collection services are often provided by local government authorities, or by private companies for industrial and commercial waste. Some areas, especially those in less developed countries, do not have formal waste-collection systems.

Waste Handling Practices

Curbside collection is the most common method of disposal in most European countries, Canada, New Zealand and many other parts of the developed world in which waste is collected at regular intervals by specialised trucks. This is often associated with curb-side waste segregation. In rural areas waste may need to be taken to a transfer station. Waste collected is then transported to an appropriate disposal facility. In some areas, vacuum collection is used in which waste is transported from the home or commercial premises by vacuum along small bore tubes. Systems are in use in Europe and North America.

Pyrolysis is used for disposal of some wastes including tires, a process that can produce recovered fuels, steel and heat. In some cases tires can provide the feedstock for cement manufacture. Such systems are used in USA, California, Australia, Greece, Mexico, the United Kingdom and in Israel. The RESEM pyrolysis plant that has been operational at Texas USA since December 2011, and processes up to 60 tons per day. In some jurisdictions unsegregated waste is collected at the curb-side or from waste transfer stations and then sorted into recyclables and unusable waste. Such systems are capable of sorting large volumes of solid waste, salvaging recyclables, and turning the rest into bio-gas and soil conditioner. In San Francisco, the local government established its Mandatory Recycling and Composting Ordinance in support of its goal of zero waste by 2020, requiring everyone in the city to keep recyclables and compostables out of the landfill. The three streams are collected with the curbsideFantastic 3 bin system – blue for recyclables, green for compostables, and black for landfill-bound materials – provided to residents and businesses and serviced by San Francisco's sole refuse hauler, Recology. The City'sPay-As-You-Throw system charges customers by the volume of landfill-bound materials, which provides a financial incentive to separate recyclables and compostables from other discards. The City's Department of the Environment's Zero Waste Program has led the City to achieve 80% diversion, the highest diversion rate in North America. Other businesses such as Waste Industries use a variety of colors to distinguish between trash and recycling cans.

Financial Models

In most developed countries, domestic waste disposal is funded from a national or local tax which may be related to income, or notional house value. Commercial and industrial waste disposal is typically charged for as a commercial service, often as an integrated charge which includes disposal costs. This practice may encourage disposal contractors to opt for the cheapest disposal option such as landfill rather than the environmentally best solution such as re-use and recycling. In some areas such as Taipei, the city government charges its households and industries for the volume of rubbish they produce. Waste will only be collected by the city council if waste is disposed in government issued rubbish bags. This policy has successfully reduced the amount of waste the city produces and increased the recycling rate.

Disposal Solutions

Landfill

A landfill compaction vehicle in action.

Spittelau incineration plant in Vienna.

Incineration

Incineration is a disposal method in which solid organic wastes are subjected to combustion so as to convert them into residue and gaseous products. This method is useful for disposal of residue of both solid waste management and solid residue from waste water management. This process reduces the volumes of solid waste to 20 to 30 percent of the original volume. Incineration and other high temperature waste treatment systems are sometimes described as "thermal treatment". Incinerators convert waste materials into heat, gas, steam, and ash.

Incineration is carried out both on a small scale by individuals and on a large scale by industry. It is used to dispose of solid, liquid and gaseous waste. It is recognized as a practical method of disposing of certain hazardous waste materials (such as biological medical waste). Incineration is a controversial method of waste disposal, due to issues such as emission of gaseous pollutants.

Incineration is common in countries such as Japan where land is more scarce, as these facilities generally do not require as much area as landfills. Waste-to-energy (WtE) or energy-from-waste (EfW) are broad terms for facilities that burn waste in a furnace or boiler to generate heat, steam or electricity. Combustion in an incinerator is not always perfect and there have been concerns about pollutants in gaseous emissions from incinerator stacks. Particular concern has focused on some very persistent organic compounds such as dioxins, furans, and PAHs, which may be created and which may have serious environmental consequences.

Recycling

Waste not the Waste. Sign in Tamil Nadu, India.

Steel crushed and baled for recycling.

Recycling is a resource recovery practice that refers to the collection and reuse of waste materials such as empty beverage containers. The materials from which the items are made can be reprocessed into new products. Material for recycling may be collected separately from general waste using dedicated bins and collection vehicles, a procedure called kerbside collection. In some communities, the owner of the waste is required to separate the materials into different bins (e.g. for paper, plastics, metals) prior to its collection. In other communities, all recyclable materials are placed in a single bin for collection, and the sorting is handled later at a central facility. The latter method is known assingle-stream recycling.

The most common consumer products recycled include aluminium such as beverages cans, copper such as wire, steel from food and aerosol cans, old steel furnishings or equipment, rubber tyres, polyethylene and PET bottles, glass bottles and jars, paperboard cartons, newspapers, magazines and light paper, and corrugated fiberboard boxes.

PVC, LDPE, PP, and PS (see resin identification code) are also recyclable. These items are usually composed of a single type of material, making them relatively easy to recycle into new products. The recycling of complex products (such as computers and electronic equipment) is more difficult, due to the additional dismantling and separation required.

The type of material accepted for recycling varies by city and country. Each city and country has different recycling programs in place that can handle the various types of recyclable materials. However, certain variation in acceptance is reflected in the resale value of the material once it is reprocessed.

Re-Use

Biological Reprocessing

An active compost heap.

Recoverable materials that are organic in nature, such as plant material, food scraps, and paper products, can be recovered through composting and digestion processes to decompose the organic matter. The resulting organic material is then recycled as mulch or compost for agricultural or landscaping purposes. In addition, waste gas from the process (such as methane) can be captured and used for generating electricity and heat (CHP/cogeneration) maximising efficiencies. The intention of biological processing in waste management is to control and accelerate the natural process of decomposition of organic matter. (See resource recovery).

Energy Recovery

Energy recovery from waste is the conversion of non-recyclable waste materials into usable heat, electricity, or fuel through a variety of processes, including combustion, gasification, pyrolyzation, anaerobic digestion, and landfill gas recovery.This process is often called waste-to-energy. Energy recovery from waste is part of the non-hazardous waste management hierarchy. Using energy recovery to convert non-recyclable waste materials into electricity and heat, generates a renewable energy source and can reduce carbon emissions by offsetting the need for energy from fossil sources as well as reduce methane generation from landfills. Globally, waste-to-energy accounts for 16% of waste management.

The energy content of waste products can be harnessed directly by using them as a direct combustion fuel, or indirectly by processing them into another type of fuel. Thermal treatment ranges from using waste as a fuel source for cooking or heating and the use of the gas fuel (see above), to fuel for boilers to generate steam and electricity in a turbine. Pyrolysis and gasification are two related forms of thermal treatment where waste materials are heated to high temperatures with limited oxygen availability. The process usually occurs in a sealed vessel under high pressure. Pyrolysis of solid waste converts the material into solid, liquid and gas products. The liquid and gas can be burnt to produce energy or refined into other chemical products (chemical refinery). The solid residue (char) can be further refined into products such as activated carbon. Gasification and advanced Plasma arc gasification are used to convert organic materials directly into a synthetic gas (syngas) composed of carbon monoxide and hydrogen. The gas is then burnt to produce electricity and steam. An alternative to pyrolysis is high temperature and pressure supercritical water decomposition (hydrothermal monophasic oxidation).

Pyrolysis

Pyrolysis is a process of thermo-chemical decomposition of organic materials by heat in the absence of oxygen which produces various hydrocarbon gases.During pyrolysis, the molecules of object are subjected to very high temperatures leading to very high vibrations. Therefore, every molecule in the object is stretched and shaken to an extent that molecules starts breaking down. The rate of pyrolysis increases with temperature. In industrial applications, temperatures are above 430 °C (800 °F). Fast pyrolysis produces liquid fuel for feedstocks like wood. Slow pyrolysis produces gases and solid charcoal. Pyrolysis hold promise for conversion of waste biomass into useful liquid fuel. Pyrolysis of waste plastics can produce millions of litres of fuel. Solid products of this process contain metals, glass, sand and pyrolysis coke which cannot be converted to gas in the process.

Resource Recovery

Resource recovery is the systematic diversion of waste, which was intended for disposal, for a specific next use. It is the processing of recyclables to extract or recover materials and resources, or convert to energy. These activities are performed at a resource recovery facility. Resource recovery is not only environmentally important, but it is also cost effective. It decreases the amount of waste for disposal, saves space in landfills, and conserves natural resources.

Resource recovery (as opposed to waste management) uses LCA (life cycle analysis) attempts to offer alternatives to waste management. For mixed MSW (Municipal Solid Waste) a number of broad studies have indicated that administration, source separation and collection followed by reuse and recycling of the non-organic fraction and energy and compost/fertilizer production of the organic material via anaerobic digestion to be the favoured path.

As an example of how resource recycling can be beneficial, many of the items thrown away contain precious metals which can be recycled to create a profit, such as the components in circuit boards. Other industries can also benefit from resource recycling with the wood chippings in pallets and other packaging materials being passed onto sectors such as the horticultural profession. In this instance, workers can use the recycled chips to create paths, walkways, or arena surfaces.

Sustainability

The management of waste is a key component in a business' ability to maintaining ISO14001 accreditation. Companies are encouraged to improve their environmental efficiencies each year by eliminating waste through resource recovery practices, which are sustainability-related activities. One way to do this is by shifting away from waste management to resource recovery practices like recycling materials such as glass, food scraps, paper and cardboard, plastic bottles and metal. This topic was on the agenda of the international Conference on Green Urbanism, held in Italy 12–14 October 2016.

Avoidance and Reduction Methods

An important method of waste management is the prevention of waste material being created, also known as waste reduction. Methods of avoidance include reuse of second-hand products, repairing broken items instead of buying new, designing products to be refillable or reusable (such as cotton instead of plastic shopping bags), encouraging consumers to avoid using disposable products (such as disposable cutlery), removing any food/liquid remains from cans and packaging, and designing products that use less material to achieve the same purpose (for example, lightweighting of beverage cans).

International Waste Movement

While waste transport within a given country falls under national regulations, trans-boundary movement of waste is often subject to international treaties. A major concern to many countries in the world has been hazardous waste. The Basel Convention, ratified by 172 countries, deprecates movement of hazardous waste from developed to less developed countries. The provisions of the Basel convention have been integrated into the EU waste shipment regulation. Nuclear waste, although considered hazardous, does not fall under the jurisdiction of the Basel Convention.

Benefits

Waste is not something that should be discarded or disposed of with no regard for future use. It can be a valuable resource if addressed correctly, through policy and practice. With rational and consistent waste management practices there is an opportunity to reap a range of benefits. Those benefits include:

- Economic – Improving economic efficiency through the means of resource use, treatment and disposal and creating markets for recycles can lead to efficient practices in the production and consumption of products and materials resulting in valuable materials being recovered for reuse and the potential for new jobs and new business opportunities.
- Social – By reducing adverse impacts on health by proper waste management practices, the resulting consequences are more appealing settlements. Better social advantages can lead to new sources of employment and potentially lifting communities out of poverty especially in some of the developing poorer countries and cities.
- Environmental – Reducing or eliminating adverse impacts on the environmental through reducing, reusing and recycling, and minimizing resource extraction can provide improved air and water quality and help in the reduction of greenhouse gas emissions.
- Inter-generational Equity – Following effective waste management practices can provide subsequent generations a more robust economy, a fairer and more inclusive society and a cleaner environment.

Challenges in Developing Countries

Waste management in cities with developing economies and economies in transition experience exhausted waste collection services, inadequately managed and uncontrolled dumpsites and the problems are worsening. Problems with governance also complicate the situation. Waste management, in these countries and cities, is an ongoing challenge and many struggle due to weak institutions, chronic under-resourcing and rapid urbanization. All of these challenges along with the lack of understanding of different factors that contribute to the hierarchy of waste management, affect the treatment of waste.

Technologies

Traditionally the waste management industry has been a late adopter of new technologies such as RFID (Radio Frequency Identification) tags, GPS and integrated software packages which enable better quality data to be collected without the use of estimation or manual data entry.

REFERENCES

1. Glossary of Environment Statistics: Series F, No. 67 / Department for Economic and Social Information and Policy Analysis, United Nations. New York: UN, 1997.
2. Editorial Board/Aims & Scope. Waste Management. 34 (3): IFC. March 2014. doi:10.1016/S0956-053X(14)00026-9.
3. Davidson, Gary (June 2011).Waste Management Practices: Literature Review (PDF). Dalhousie University - Office of Sustainability.
4. Guidelines for National Waste Management Strategies Moving from Challenges to Opportunities (PDF). United Nations Environmental Programme. 2013. ISBN 978-92-807-3333-4..
5. Barbalace, Roberta Crowell (2003-08-01).The History of Waste. EnvironmentalChemistry.com. Retrieved 2013-12-09.
6. Florence Nightingale, Selected Writings of Florence Nightingale, cd. Lucy Ridgely Seymer (New York: The Macmillan Co., 1954), pp. 38287
7. Herbert, Lewis (2007). "Centenary History of Waste and Waste Managers in London and South East England". Chartered Institution of Wastes Management.
8. Chadwick, Edwin (1842). Report...from the Poor Law Commissioners on an Inquiry into the Sanitary Conditions of the Labouring Population of Great Britain. London. pp. 369–372. via Laura Del Col (11 October 2002).Chadwick's Report on Sanitary Conditions. The Victorian Web.
9. National Waste & Recycling Association.History of Solid Waste Management. Washington, D.C. Retrieved 2013-12-09.
10. Gandy, Matthew (1994). Recycling and the Politics of Urban Waste. Earthscan. ISBN 9781853831683.
11. Covered Bodies.
12. RESEM Waste Tyre Pyrolysis Plant in USA, retrieved 2011-10-24
13. http://www.siemens.com/entry/cc/features/greencityindex_international/all/en/pdf/report_northamerica_en.pdf
14. City of Chicago, Illinois. Department of Streets and Sanitation. "What is Single Stream Recycling. Accessed 2013-12-09.
15. Montgomery County, Maryland. Division of Solid Waste Services. "Curbside Collection." Accessed 2013-12-09.
16. Energy Recovery from Waste. USEPA. 2014.

17. Waste Hierarchy. New Energy Corporation. 2014.
18. Oxford Reference – Pyrolysis
19. Encyclopedia Britannica
20. By Prabir Basu: Biomass Gasification, Pyrolysis and Torrefaction: Practical Design and Theory
21. Frequent Questions. USEPA. 2012.
22. Removing food remains to reduce waste. Recycling Guide. Retrieved 2012-09-25.
23. Schneider, Michael; Johnson, Liz.Lightweighting. Projects in Scientific Computing. Pittsburgh Supercomputing Center, Carnegie Mellon University, University of Pittsburgh. Retrieved 2012-09-25.
24. Waste Management. Science Direct. 33 (1): 220–232. 2013. Missing or empty |title= (help)
25. Claire Swedberg (4 February 2014).Air-Trak Brings Visibility to Waste Management. RFID Journal. Retrieved 1 October 2015.

CHAPTER 8

Waste Collection and Management

Chapter Contents

- **Introduction**
- **Household Waste**
- **Commercial Waste**
- **Employment as Garbage Collector**
- **Waste Sorting**
- **Waste-To-Energy**

Our general waste collection service takes care of all non-recyclable waste produced by your business including low grade plastics, polythene & unrecyclable packaging.

Many businesses often combine our general waste service with a recycling collection. This ensures that your waste is segregated effectively, helping you to meet current waste legislation. Whatever the size of your business, we have a waste collection package that is right for you.

We assess your bin size requirements, the number of bins required, the number of business locations, the type of waste and the number of collections needed, to ensure the package is based on the individual needs of your business.

INTRODUCTION

Waste collection is a part of the process of waste management. It is the transfer of solid waste from the point of use and disposal to the point of treatment or landfill. Waste collection also includes the curbside collection of recyclable materials that technically are not waste, as part of a municipal landfill diversion program.

A waste collection vehicle in Sakon Nakhon, Thailand.

Manual waste collection in Bukit Batok West, Singapore.

Waste on a sidewalk for collection, bagged and stickered - in Dublin, Ireland.

HOUSEHOLD WASTE

Household waste in economically developed countries will generally be left in waste containers or recycling bins prior to collection by a waste collector using a waste collection vehicle.

However, in many developing countries, such as Mexico and Egypt, waste left in bins or bags at the side of the road will not be removed unless residents interact with the waste collectors.

Mexico City residents must haul their trash to a waste collection vehicle which makes frequent stops around each neighborhood. The waste collectors will indicate their readiness by ringing a distinctive bell and possibly shouting. Residents line up and hand their trash container to the waste collector. A tip may be expected in some neighborhoods. Private contracted waste collectors may circulate in the same neighborhoods as many as five times per day, pushing a cart with a waste container, ringing a bell and shouting to announce their presence. These private contractors are not paid a salary, and survive only on the tips they receive. Later, they meet up with a waste collection vehicle to deposit their accumulated waste.

The waste collection vehicle will often take the waste to a transfer station where it will be loaded up into a larger vehicle and sent to either a landfill or alternative waste treatment facility.

COMMERCIAL WASTE

Waste collection considerations include type and size of bins, positioning of the bins, and how often bins are to be serviced. Overfilled bins result in rubbish falling out while being tipped. Hazardous rubbish like empty petrol cans can cause fires igniting other trash when the truck compactor is operating. Bins may be locked or stored in secure areas to avoid having non-paying parties placing rubbish in the bin.

EMPLOYMENT AS GARBAGE COLLECTOR

The Bureau of Labor Statistics predicts that employment in this industry will increase 16.2 percent, adding 21,600 new jobs, by 2022. The main drivers of this growth include a rise in population, individual income and more people choosing to recycle.

Waste Picker

Scavenging in Jakarta, Indonesia.

A **waste picker** is a person who salvages reusable or recyclable materials thrown away by others to sell or for personal consumption. There are millions of waste pickers worldwide, predominantly in developing countries, but increasingly in post-industrial countries as well.

Forms of waste picking have been practiced since antiquity, but modern traditions of waste picking took root during industrialization in the nineteenth century. Over the past half-century, waste picking has expanded vastly in the developing world due to urbanization, toxic colonialism and the global waste trade. Many cities only provide solid waste collection.

Terminology

Many terms are used to refer to people who salvage recyclables from the waste stream for sale or personal consumption. In English, these terms include *rag picker*, *reclaimer*, *informal resource recoverer*, *binner*, *recycler*, *poacher*, *salvager*, *scavenger*, and *waste picker*; in Spanish *cartonero*, *chatarrero*, *pepenador*, *clasificador*, *minador* and *reciclador*; and in Portuguese *catador de materiais recicláveis*. A more contemporary term, focusing on the outcome of the professional activity, is "informal sector recycling". However, the word "informal" can be partly misleading, because in practice a continuum between total informality and proper organization in taxed registered formal activities may be encountered.

In 2008, participants of the The First World Conference of Waste Pickers chose to use the term "waste picker" for English usage to facilitate global communication. The term "scavenger" is also commonly used, but many waste pickers find it demeaning due to the implied comparison with animals.

A waste picker is different from a waste collector because the waste collected by the latter may be destined for a landfill or incinerator, not necessarily for a recycling facility.

Dumpster diving generally refers to the practice of anti-consumer and freegan activists who reclaim items such as food and clothes from the waste stream as a form of protest against consumer culture. "Waste picking" generally refers to activity motivated purely by economic need.

Prevalence and Demographics

There is little reliable data about the number and demographics of waste pickers worldwide. Most academic research on waste pickers is qualitative rather than quantitative. Systematic large-scale data collection is difficult due to the profession's informal nature, porous borders, seasonally fluctuating workforce, and widely dispersed and mobile worksites. Also, many researchers are reluctant to produce quantitative data out of fear that it might be used to justify crackdowns on waste picking by authorities. Thus, the large-scale estimates that do exist are mainly extrapolations based on very small original research samples. In his book, "The World's Scavengers" (2007), Martin Medina provides a methodological guide to researching waste picking.

In 1988, the World Bank estimated that 1–2% of the global population subsists by waste picking. A 2010 study estimates that there are 1.5 million waste pickers in India alone. Brazil, the country that collects the most robust official statistics on waste pickers, estimates that nearly a quarter million of its citizens engage in waste picking.

Waste picker incomes vary vastly by location, form of work, and gender. Some waste pickers live in extreme poverty, but many others earn multiple times their country's minimum wage. Recent studies indicate that waste pickers in Belgrade, Serbia, earn approximately US$3 per day, while waste pickers in Cambodia typically earn $1 per day. Official statistics in Brazil indicate that men earn more than women, regardless of age. Approximately two thirds of Brazil's waste pickers are men overall, but this proportion jumps to 98% in high income waste picker groups (those earning between 3–4 times the minimum wage). No women were found in the highest income groups (those earning over 10 times the minimum wage).

Causes

In Developing Countries

Over the past half century, in-country migration and increased fertility rates have caused the population of cities in the developing world to mushroom. The global population of urban dwellers is expected to double between 1987 and 2015, with 90% of this growth occurring in developing countries. Much of the new population has settled in urban slums and squatter settlements, which have expanded rapidly with no central planning. The United Nations Habitat Report found that nearly one billion people worldwide live in slums, about a third of the world's urban dwellers.

The rapid urbanization greatly increased the demand for informal waste collecting services, as cities lacked the infrastructure and resources to collect the totality of wastes generated by their inhabitants. Despite spending 30–50% of operation budgets on waste management, developing world cities today collect only 50–80% of refuse generated by inhabitants. Residents and businesses often resort to burning garbage or disposing of it in streets, rivers, vacant lots, and open dumps. This is a source of air, land, and water pollution that threatens human health and the environment. Informal waste collectors help mitigate this harm by collecting recyclable materials by foot or in pushcarts, tricycles, donkey carts, horse carts, and pickup trucks.

On the supply side, urbanization has facilitated the expansion of waste picking by creating a large pool of unemployed and underemployed residents with few alternative means of earning a livelihood. Known as "the one industry that is always hiring," waste picking provides a cushion for many who lose their jobs during times of war, crisis, and economic downturn in countries that do not have welfare systems. It is also one of the few work opportunities available to people who lack formal education or job experience.

In post-Industrial Countries

Though documentation exists of rag pickers and scrap metal collectors supplying goods to paper mills and foundries as early as the 17th century, modern waste picking did not flourish in the US and Europe until the 19th century. Just as in the developing world, the combination of industrialization and urbanization led to three trends which favored the blossoming of the informal waste collecting industry: increased generation of urban waste, increased demand for raw materials from industry, and increased numbers of urban dwellers in need of livelihoods. In that era, waste pickers were known as wharf rats, tinkers, rag and bone men, mudlarks, and ragpickers. By the mid-20th century waste picking decreased, as waste management industries were formalized, and welfare states decreased the poor's reliance on informal recycling.

Beginning in the mid-1990s, however, informal recycling in parts of the United States and Western Europe once again began to mushroom. Two factors fueled the boom: First, the demand for recycling surged due to the increased waste stream, declining room in landfills, new recycling technologies, and the efforts of environmentalists. In 1985 only one roadside recycling program existed in the United States. By 1998, there were 9,000 such programs and 12,000 recyclable drop-off centers. Laws were passed in some states making it illegal not to recycle. Second, changes in the political economy including the loss of manufacturing jobs, cutbacks to government employment, and the roll back of the welfare state increased the ranks of the poor, working poor, and homeless—thus there were more people disposed to wastepick as a full-time profession or supplemental job.

American waste pickers predominantly collect cans, bottles, and cardboard. Many immigrants work as waste pickers because language and documentation barriers limit their opportunities to work elsewhere. Many homeless people also work as waste pickers—some describe it as their only alternative to panhandling.Some recyclers use vans to increase their yield while others work on foot with carts. Anecdotal evidence suggests that most American waste pickers are male, as waste picking is widely considered too dirty and strenuous a job for women. During an ethnography of homeless recyclers in San Francisco, sociologist Teresa Gowan claims to have met hundreds of male waste pickers, but only four female waste pickers.

Costs and Benefits

Social and Ecological Benefits

A scavenger in Hong Kong pours water onto the paper she has collected in order to increase her profit by adding to its weight.

Waste picking offers significant ecological, economic, and social benefits:

- Job creation: Waste picking provides a source of livelihood to extremely poor people with few other employment opportunities. Though many waste pickers practice their trade as a full-time profession, its flexible hours make it accessible to women with other care responsibilities and to people looking to supplement income from other jobs. During times of need, waste picking serves as a safety net to street children, orphans, the elderly, widows, migrants, the disabled, the unemployed, and victims of armed conflicts. Waste picking also benefits the broader economy by supplying raw materials to industry and creating many associated jobs for middlemen who purchase, sort, process, and resell materials collected by waste pickers.
- Public health and sanitation: Waste pickers collect garbage from neighborhoods that lack public services. Without waste pickers, residents would be forced to burn trash, or dispose of it in rivers, streets and empty lots. Waste pickers provide the only solid waste removal service in many cities.

- Municipal savings: Waste pickers provide between 50 and 100% of waste collecting services in most cities of the developing world, according to a 2010 UN Habitat report. This is effectively serves as a mass subsidy for city governments, who do not pay for the labor. Moreover, recycling expands the lifespan of city dumps and landfills.
- Reducing pollution and mitigating climate change: By cutting the quantity of virgin materials needed for production, waste pickers save room in landfills, lessen water and energy consumption, reduce air and water pollution, and abate climate change. Since 2009, international delegations of waste pickers have attended at least five global climate change conferences to demand that climate funds invest in resource recovery programs that will help ensure waste pickers' livelihoods, rather than waste disposal technologies like incinerators.

Social Costs

Waste pickers not only generate social benefits, but also potential costs as well. These include:

- Occupational hazards: See discussion below.
- Child labor: Children commonly work as waste pickers. This may interfere with their education, or harm their physical, emotional, and social well-being.
- Litter: Waste pickers working on streets sometimes spread waste from trash bags, sullying the sidewalk and creating more work for city street sweepers.
- Public nuisance: Many people view waste pickers as a nuisance or source of shame for their communities. Waste pickers' perceived poverty and lack of sanitation makes some people uncomfortable or fearful. In developing countries especially, many argue that modern services should replace waste pickers.
- Pilfering of public property: In some cities, waste pickers have been known to steal, meltdown, and resell public property such as telephone electrical copper wires, steel fence, or manhole coverings.

Occupational Hazards

- Health risks: There is a high prevalence of infectious disease among waste pickers due to their exposure to hazardous materials such as fecal matter, paper saturated by toxic materials, bottles and containers with chemical residues, contaminated needles, and heavy metals from batteries. A study in Mexico City found the average lifespan of a dumpsite waste collector to be 39 years, compared to the national average of 69 years, though a later World Bank study estimated life span to be 53 years. In Port Said, Egypt, a 1981 study showed an infant mortality rate of 1/3 among waste pickers (one out of three babies dies before reaching age one).
- Risks of injury: Among the most common types of job-related injuries for waste pickers are back and hand injuries caused by lifting heavy objects with

little equipment. In a study of 48 waste pickers in Santo André, Brazil, almost all workers reported pain in the back, legs, shoulder, arms, and hands. Waste pickers who work in open dumps are exposed to large amounts of toxic fumes, and face other severe threats including being run over by trucks and caught in surface subsidence, trash slides, and fires. On 10 July 2000, several hundred waste pickers were killed by a trash slide from a huge garbage mountain after monsoon rains at an open dump in Payatas, Philippines.

- Stigma, harassment, and violence: Most waste picking activity is illegal or unpermitted, so waste pickers commonly face harassment by police and authorities. Also, there is widespread public scorn against waste pickers due to their poverty and perceived lack of hygiene. One of the most extreme manifestations of such stigma occurs in Colombia, where, since the 1980s, social cleansing vigilante groups, sometimes working with police complicity, have killed at least two thousand waste collectors, beggars, and prostitutes—whom they refer to as disposables (*desechables*). In 1992, around the peak of this activity, eleven corpses of murdered waste collectors were discovered at a university in Barranquilla. Their organs had been sold for transplants and bodies sold to the medical school for dissection (Medina 2009, 155).

Waste Picker Organizing

Traditionally, scholars assumed that informal workers such as waste pickers could not collectively organize due to structural barriers such as lack of legal protection, widely dispersed worksites, porous borders to their profession, a culture of independence and individualism, lack of institutional experience, and lack of money and time to build organizations. Nonetheless, in recent decades waste pickers across Latin America, Asia, and Africa have begun collectively organizing to win a place within formal recycling systems.

Waste pickers use many organizational formats including cooperatives, associations, companies, unions, and micro-enterprises. Despite the differences in format, most of these organizations share three primary purposes. First, by pooling capital, establishing microenterprises, and forming partnerships with business and government, waste collectors increase their selling power. Second, by securing uniforms, safety equipment, and work permits, cooperatives increase workplace dignity and safety. And third, by demanding recognition and compensation from the state for their environmental and economic contributions, cooperatives increase members' political might. These three functions—political influence, workplace dignity and safety, and increased earnings—are mutually reinforcing, like legs on a stool.

Some waste pickers have created "women's only" organizations, which seek to combat gender-based discrimination at worksites and in communities. A study in Brazil indicates that women are heavily overrepresented even in coed organizations, making up 56% of the membership despite the fact that they represent only a third of the total waste picking population.

Beginning in the 1990s, waste picker organizations in many parts of the world began uniting into regional, national, and transnational coalitions to increase their political voice and economic leverage. In March 2008, delegates from 30 countries gathered in Bogotá, Colombia, for the first World Conference (and Third Latin American Conference) of Waste Pickers (WIEGO 2008). One of the key issues discussed was the global trend of privatization and concentration of waste management systems. Normally, privatization is thought of as the handover of government functions to the private sector, but in this case, privatization often means the transference of services formerly provided by informal waste collectors to private firms. As waste streams have expanded due to increased consumption, the waste management industry is becoming increasingly lucrative. Governments around the world are granting private companies monopolies on waste management systems, meaning that the cooperatives' survival hinges on building political and economic alliances needed to win contracts—often an uphill battle given authorities' distrust of waste collectors and the cooperatives' lack of capital for modern machinery.

Organizing in Latin America

In 1962, the first known Latin American waste picker's organization, *Cooperativa Antioqueña de Recolectores de Subproductos*, was created in Medellín, Colombia. The Colombian waste pickers movement did not emerge as a veritable political force until 1990, however, when four cooperatives who been fighting the closure of a landfill united as the Waste Collector's Association of Bogotá (ARB). Today the ARB is one of the world's most active and established waste picker organizations. Throughout the 1990s, powerful waste collectors associations began to form in other Latin American countries as well—most notably in Brazil, Argentina, Chile, and Uruguay.

In 2005, Brazil hosted the first meeting of the Latin American Waste Picker Network (LAWPN)—an organization that now represents waste pickers movements from 16 countries. LAWPN has four key functions. First, it facilitates exchanges of knowledge, technology, and strategies between member organizations through regional conventions, country-to-country delegations, telecommunications, and strategic reports. Second, it organizes transnational solidarity to aid in local battles. For example, when waste pickers in Montevideo, Uruguay, needed support in a local campaign, member organizations across Latin America issued solidarity statements and pressured their national ambassadors in Uruguay to do the same. Third, LAWPN sends leaders from countries with strong waste picker movements to countries with weak movements in order to promote the development of new leadership and organizations. Fourth, LAWPN organizes global waste picker committees to make appeals for support to transnational governance organizations such as the Inter-American Development Bank, the UN Convention on Climate Change, and the International Labour Organization.

In Argentina, The Movement of Excluded Workers (excluded in the sense that their work is not recognized by the government and they are excluded from receiving rights) is the largest waste pickers organization. It is a social organization, independent from political parties, which brings together more than 2,000 cartoneros (waste picker) in Capital Federal and the suburbs, specifically in the neighborhoods of Lanús and Lomas de Zamora. After years of sacrifice and struggle, they managed to improve their working conditions. They have established a more logistical system—they no longer travel by hanging off of trucks, they have obtained a work incentive and uniforms, and lastly they have founded a nursery for 160 children, some of whom worked in the past as cartoneros. However, they still have to advance in designing a social security program that includes all the cartoneros of Buenos Aires. Additionally, they have to raise awareness among residents of the city to separate their trash so that they can collect door to door, without having direct contact with wet materials.

Organizing in Asia

India is home to Asia's largest waste picker movement. Self-Employed Women's Association of India, a trade union that exclusively organizes women in the informal economy and has membership of over one million, began organizing waste pickers in the late 1970s. SEWA has created nearly ninety waste pickers cooperatives, which members use to gain collective work contracts and access to credit, training, and markets. One of Asia's largest waste picker associations is The Alliance of Indian Waste Pickers (AIW), a national network of 35 organizations in 22 cities. The AIW facilitates peers support, advocacy and cross-learning among waste pickers, iterant buyers, and support NGOs. Also in India, the All India Kabari Mazdoor Mahasangh (AIKMM) is locked in a battle with the New Delhi Municipal Council, which closed a deal with the Hyderabad-based company, Ramky Energy and Environment Ltd to manage the waste, in effect criminalising the work of more than 100,000 unorganised waste pickers that currently sort about 20% of Delhi's garbage. The EJOLT project made a video on what they call the Delhi waste war. Waste pickers in Thailand have also organized, including the Khon Kaen scavengers.

An Indian waste pickers union known as KKPKP recently carried out a mapping initiative to identify organizations of or that work with waste pickers across the continent—a first step towards the development an Asian network. Several NGOs and trade unions that work with waste pickers, as well as loosely formed groups of waste pickers, were identified in Cambodia, Indonesia, The Philippines, and Thailand.

In Bangladesh, Grambangla Unnayan Committee is working with the waste picker community at Matuail Waste Dumpsite of Dhaka City Corporation. A daycare centre and a non-formal primary school has been established for the child waste pickers where 112 child waste pickers are given early childhood care and education. Women waste pickers of Matuail has formed a Waste Pickers' cooperative.

Organizing in Africa

Egypt has of the world's most well-established and robust informal recycling systems. The labor is done for the most part by the Zabaleen (informal waste collectors), a predominantly Coptic Christian community, which in the 1940s began collecting garbage—work viewed as impure by Egypt's Muslim majority. In 2003, the Zabaleen's existence and way of life came under threat when Cairo authorities awarded annual contracts $50 to three multinational garbage disposal companies, pushing the Zabaleen to collectively defend their livelihood.

The South African Waste Picker organization held its first meeting in July 2009, which included 100 waste pickers from 26 landfills across the country. There are no present plans to create an African waste picker's network, but Shack/Slum Dwellers International organized meetings between waste picker leaders in Kenya, Egypt, and South Africa.

WASTE SORTING

Recycling bins in Singapore.

Manual waste sorting for recycling.

Characteristic containers for recycling in Portovenere, Italy.

Garbage containers in Fuchū, Tokyo, Japan.

Waste sorting is the process by which waste is separated into different elements. Waste sorting can occur manually at the household and collected through curbside collection schemes, or automatically separated in materials recovery facilities or mechanical biological treatment systems. Hand sorting was the first method used in the history of waste sorting.

Waste can also be sorted in a civic amenity site.

Waste segregation means dividing waste into dry and wet. **Dry waste** includes wood and related products, metals and glass. **Wet waste**, typically refers to organic waste usually generated by eating establishments and are heavy in weight due to dampness. Waste can also be segregated on basis of biodegradable or non-biodegradable waste.

Landfills are an increasingly pressing problem. Less and less land is available to deposit refuse, but the volume of waste is growing all time. As a result, segregating waste is not just of environmental importance, but of economic concern, too.

Methods

Waste is collected at its source in each area and separated. The way that waste is sorted must reflect local disposal systems. The following categories are common:

- Paper
- Cardboard (including packaging for return to suppliers)
- Glass (clear, tinted – no light bulbs or window panes, which belong with residual waste)
- Plastics
- Textiles
- Wood, leather, rubber
- Scrap metal

- Compost
- Special/hazardous waste
- Residual waste

Organic waste can also be segregated for disposal:

- Leftover food which has had any contact with meat can be collected separately to prevent the spread of bacteria.
- Meat and bone can be retrieved by bodies responsible for animal waste
- If other leftovers are sent, for example, to local farmers, they can be sterilised before being fed to the animals
- Peel and scrapings from fruit and vegetables can be composted along with other degradable matter. Other waste can be included for composting, too, such as cut flowers, corks, coffee grindings, rotting fruit, tea bags, egg- and nutshells, paper towels etc.

Chip pan oil (fryer oil), used fats, vegetable oil and the content of fat filters can be collected by companies able to re-use them. Local authority waste departments can provide relevant addresses. This can be achieved by providing recycling bins.

By Country

In Germany, regulations exist that provide mandatory quotas for the waste sorting of packaging waste and recyclable materials such as glass bottles.

In Denpasar, Bali, Indonesia, a pilot project using an automated collecting machine of plastic bottles or aluminium cans with voucher reward has been implemented in a market.

WASTE-TO-ENERGY

Spittelau incineration plant is one of several plants that provide district heating in Vienna.

Waste-to-energy (WtE) or **energy-from-waste** (EfW) is the process of generating energy in the form of electricity and/or heat from the primary treatment of waste. WtE is a form of energy recovery. Most WtE processes produce electricity and/or heat directly through combustion, or produce a combustible fuel commodity, such as methane, methanol, ethanol or synthetic fuels.

The first incinerator or "Destructor" was built in Nottingham UK in 1874 by Manlove, Alliott & Co. Ltd. to the design of Albert Fryer.

- The first US incinerator was built in 1885 on Governors Island in New York, NY.
- The first waste incinerator in Denmark was built in 1903 in Frederiksberg
- The first facility in Czech Republic was built in 1905 in Brno.

Incineration

Incineration, the combustion of organic material such as waste with energy recovery, is the most common WtE implementation. All new WtE plants in OECD countries incinerating waste (residual MSW, commercial, industrial or RDF) must meet strict emission standards, including those on nitrogen oxides (NO_x), sulphur dioxide (SO_2), heavy metals and dioxins. Hence, modern incineration plants are vastly different from old types, some of which neither recovered energy nor materials. Modern incinerators reduce the volume of the original waste by 95-96 percent, depending upon composition and degree of recovery of materials such as metals from the ash for recycling.

Incinerators may emit fine particulate, heavy metals, trace dioxin and acid gas, even though these emissions are relatively low from modern incinerators. Other concerns include proper management of residues: toxic fly ash, which must be handled in hazardous waste disposal installation as well as incinerator bottom ash (IBA), which must be reused properly.

Critics argue that incinerators destroy valuable resources and they may reduce incentives for recycling. The question, however, is an open one, as countries in Europe recycling the most (up to 70%) also incinerate their residual waste to avoid landfilling.

Incinerators have electric efficiencies of 14-28%. In order to avoid losing the rest of the energy, it can be used for e.g. district heating (cogeneration). The total efficiencies of cogeneration incinerators are typically higher than 80% (based on the lower heating value of the waste).

The method of using incineration to convert municipal solid waste (MSW) to energy is a relatively old method of WtE production. Incineration generally entails burning waste (residual MSW, commercial, industrial and RDF) to boil water which powers steam generators that make electric energy and heat to be used in homes, businesses, institutions and industries. One problem associated with incinerating MSW to make electrical energy is the potential for pollutants to enter the atmosphere with the flue gases from the boiler. These pollutants can be acidic and in the 1980s were reported to cause environmental damage by turning rain into acid rain. Since then, the industry has removed this problem by the use of lime scrubbers and electro-static precipitators on smokestacks. By passing the smoke through the basic lime scrubbers, any acids that might be in the smoke are neutralized which prevents the acid from reaching the atmosphere and hurting the environment. Many other devices, such as fabric filters, reactors, and catalysts destroy or capture other regulated pollutants. According to the New York Times, modern incineration plants are so clean that "many times more dioxin is now released from home fireplaces and backyard barbecues than from incineration. According to the German Environmental Ministry, because of stringent regulations, waste incineration plants are no longer significant in terms of emissions of dioxins, dust, and heavy metals.

WtE Technologies other than Incineration

There are a number of other new and emerging technologies that are able to produce energy from waste and other fuels without direct combustion. Many of these technologies have the potential to produce more electric power from the same amount of fuel than would be possible by direct combustion. This is mainly due to the separation of corrosive components (ash) from the converted fuel, thereby allowing higher combustion temperatures in e.g. boilers, gas turbines, internal combustion engines, fuel cells. Some are able to efficiently convert the energy into liquid or gaseous fuels:

Thermal technologies:

- Gasification: produces combustible gas, hydrogen, synthetic fuels
- Thermal depolymerization: produces synthetic crude oil, which can be further refined
- Pyrolysis: produces combustible tar/biooil and chars
- Plasma arc gasification or plasma gasification process (PGP): produces rich syngas including hydrogen and carbon monoxide usable for fuel cells or generating electricity to drive the plasma arch, usable vitrified silicate and metal ingots, salt and sulphur

Non-thermal technologies:

- Anaerobic digestion: Biogas rich in methane
- Fermentation production: examples are ethanol, lactic acid, hydrogen
- Mechanical biological treatment (MBT)
- MBT + Anaerobic digestion
- MBT to Refuse derived fuel

Global WtE Developments

During the 2001-2007 period, the WtE capacity increased by about four million metric tons per annum. Japan and China built several plants that were based on direct smelting or on fluidized bed combustion of solid waste. In China there are about 434 WtE plants in early 2016. Japan is the largest user in thermal treatment of MSW in the world with 40 million tons. Some of the newest plants use stoker technology and others use the advanced oxygen enrichment technology. There are also over one hundred thermal treatment plants using relatively novel processes such as direct smelting, the Ebara fluidization process and the Thermo- select -JFE gasification and melting technology process. In Patras, Greece, a

Greek company just finished testing a system that shows potential. It generates 25kwatts of electricity and 25kwatts of heat from waste water. In India its first energy bio-science center was developed to reduce the country's green house gases and its dependency on fossil fuel. As of June 2014, Indonesia had a total of 93.5MW installed capacity of WtE, with a pipeline of projects in different preparation phases together amounting to another 373MW of capacity.

Biofuel Energy Corporation of Denver, CO, opened two new biofuel plants in Wood River, NE, and Fairmont, MN, in July 2008. These plants use distillation to make ethanol for use in motor vehicles and other engines. Both plants are currently reported to be working at over 90% capacity. Fulcrum BioEnergy incorporated located in Pleasanton, CA, is building a WtE plant near Reno, NV. The plant is scheduled to open in 2019 under the name of Sierra BioFuels plant. BioEnergy incorporated predicts that the plant will produce approximately 10.5 million gallons per year of ethanol from nearly 200,000 tons per year of MSW.

Waste to energy technology includes fermentation, which can take biomass and create ethanol, using waste cellulosic or organic material. In the fermentation process, the sugar in the waste is changed to carbon dioxide and alcohol, in the same general process that is used to make wine. Normally fermentation occurs with no air present. Esterification can also be done using waste to energy technologies, and the result of this process is biodiesel. The cost effectiveness of esterification will depend on the feedstock being used, and all the other relevant factors such as transportation distance, amount of oil present in the feedstock, and others.Gasification and pyrolysis by now can reach gross thermal conversion efficiencies (fuel to gas) up to 75%, however a complete combustion is superior in terms of fuel conversion efficiency. Some pyrolysis processes need an outside heat source which may be supplied by the gasification process, making the combined process self-sustaining.

Carbon Dioxide Emissions

In thermal WtE technologies, nearly all of the carbon content in the waste is emitted as carbon dioxide (CO_2) to the atmosphere (when including final combustion of the products from pyrolysis and gasification; except when producing bio-char for fertilizer). Municipal solid waste (MSW) contain approximately the same mass fraction of carbon as CO_2 itself (27%), so treatment of 1 metric ton (1.1 short tons) of MSW produce approximately 1 metric ton (1.1 short tons) of CO_2.

In the event that the waste was landfilled, 1 metric ton (1.1 short tons) of MSW would produce approximately 62 cubic metres (2,200 cu ft) methane via the anaerobic decomposition of the biodegradable part of the waste. This amount of methane has more than twice the global warming potential than the 1 metric ton (1.1 short tons) of CO_2, which would have been produced by combustion. In some countries, large amounts of landfill gas are collected, but still the global warming potential of the landfill gas emitted to atmosphere in e.g. the US in 1999 was approximately 32% higher than the amount of CO_2 that would have been emitted by combustion.

In addition, nearly all biodegradable waste is biomass. That is, it has biological origin. This material has been formed by plants using atmospheric CO_2 typically within the last growing season. If these plants are regrown the CO_2 emitted from their combustion will be taken out from the atmosphere once more.

Such considerations are the main reason why several countries administrate WtE of the biomass part of waste as renewable energy. The rest—mainly plastics and other oil and gas derived products—is generally treated as non-renewables.

Determination of the Biomass Fraction

MSW to a large extent is of biological origin (biogenic), e.g. paper, cardboard, wood, cloth, food scraps. Typically half of the energy content in MSW is from biogenic material. Consequently, this energy is often recognised as renewable energy according to the waste input. Several methods have been developed by the European CEN 343 working group to determine the biomass fraction of waste fuels, such as Refuse Derived Fuel/ Solid Recovered Fuel. The initial two methods developed (CEN/TS 15440) were the **manual sorting method** and the **selective dissolution method**. A detailed systematic comparison of these two methods was published in 2010. Since each method suffered from limitations in properly characterizing the biomass fraction, two alternative methods have been developed. The first method uses the principles of radiocarbon dating. A technical review (CEN/TR 15591:2007) outlining the carbon 14 method was published in 2007. A technical standard of the carbon dating method (CEN/TS 15747:2008) will be published in 2008. In the United States, there is already an equivalent carbon 14 method under the standard method ASTM D6866.

The second method (so-called **balance method**) employs existing data on materials composition and operating conditions of the WtE plant and calculates the most probable result based on a mathematical-statistical model. Currently the balance method is installed at three Austrian and eight Danish incinerators.

A comparison between both methods carried out at three full-scale incinerators in Switzerland showed that both methods came to the same results.

Carbon 14 dating can determine with precision the biomass fraction of waste, and also determine the biomass calorific value. Determining the calorific value is important for green certificate programs such as the Renewable Obligation Certificate program in the United Kingdom. These programs award certificates based on the energy produced from biomass. Several research papers, including the one commissioned by the Renewable Energy Association in the UK, have been published that demonstrate how the carbon 14 result can be used to calculate the biomass calorific value. The UK gas and electricity markets authority, Ofgem, released a statement in 2011 accepting the use of Carbon 14 as a way to determine the biomass energy content of waste feedstock under their administration of the Renewables Obligation. Their Fuel Measurement and Sampling (FMS) questionnaire describes the information they look for when considering such proposals.

Examples of Waste-to-Energy Plants

According to ISWA there are 431 WtE plants in Europe (2005) and 89 in the United States (2004). The following are some examples of WtE plants.

Waste incineration WtE plants

- Essex County Resource Recovery Facility, Newark, New Jersey
- Lee County Solid Waste Resource Recovery Facility, Fort Myers, Florida, USA (1994)
- Montgomery County Resource Recovery Facility in Dickerson, Maryland, USA (1995)
- Spittelau (1971), and Flötzersteig (1963), Vienna, Austria (Wien Energie)
- SYSAV waste-to-energy plant in Malmö (2003 and 2008), Sweden
- Algonquin Power, Brampton, Ontario, Canada
- Stoke Incinerator, Stoke-on-Trent, UK (1989)
- Teesside EfW plant near Middlesbrough, North East England (1998)
- Edmonton Incinerator in Greater London, England (1974)
- Burnaby Waste-to-Energy Facility, Metro Vancouver, Canada (1988)
- Timarpur-Okhla Waste to Energy Plant, New Delhi, India

Liquid fuel producing plants

A single plant is currently under construction. None are yet in commercial operation:

- Edmonton Waste-to-ethanol Facility located in Edmonton, Alberta, Canada based on the Enerkem-process, fueled by RDF. Initially scheduled for completion during 2010 commissioning of front-end systems commenced December 2013 and Enerkem then expected initial methanol production during 2014.Production start has been delayed several times. As of spring 2016 Enerkem expected ethanol production to commence some time in 2017, and no public confirmation of any actual RDF processing was available.

Plasma Gasification Waste-to-Energy plants

- The US Air Force once tested a Transportable Plasma Waste to Energy System (TPWES) facility (PyroGenesis technology) at Hurlburt Field, Florida. The plant, which cost cost $7.4 million to construct, was closed and sold at a government liquidation auction in May 2013, less than three years after its commissioning. The opening bid was $25. The winning bid was sealed.

Besides large plants, domestic waste-to-energy incinerators also exist. For example, the refuge de Sarenne has a domestic waste-to-energy plant. It is made by combining a wood-fired gasification boiler with a Stirling motor.

REFERENCES

1. Herbert, Lewis (2007). "Centenary History of Waste and Waste Managers in London and South East England" (PDF). Chartered Institution of Wastes

Management.

2. Lapčík; et al. (Dec 2012). Možnosti Energetického Využití Komunálního Odpadu. GeoScience Engineering.
3. DIRECTIVE 2000/76/EC OF THE EUROPEAN PARLIAMENT AND OF THE COUNCIL of 4 December 2000 on the incineration of waste. European Union. 4 December 2000.
4. Emissionsfaktorer og emissionsopgørelse for decentral kraftvarme, Kortlægning af emissioner fra decentrale kraftvarmeværker, Ministry of the Environment of Denmark 2006 (in Danish)
5. Waste Gasification: Impacts on the Environment and Public Health
6. Environment in the EU27 Landfill still accounted for nearly 40% of municipal waste treated in the EU27 in 2010. European Union. 27 March 2012.
7. Waste–to–Energy in Austria, White Book, 2nd Edition 2010 (PDF). Austrian Ministry of Life.
8. Rosenthal, Elisabeth (12 April 2010). Europe Finds Clean Energy in Trash, but U.S. Lags. The New York Times.
9. Waste incineration – A potential danger? Bidding farewell to dioxin spouting(PDF). Federal Ministry for Environment, Nature Conservation and Nuclear Safety. September 2005.
10. Waste Council Attracts Experts Worldwide. columbia.edu. Retrieved 28 February 2015.
11. Waste to energy in Indonesia. The Carbon Trust. June 2014. Retrieved 22 July 2014.
12. Cost Effective Waste to Energy Technologies – Updated Article With Extra Information. bionomicfuel.com. Retrieved 28 February 2015.
13. The Viability of Advanced Thermal Treatment of MSW in the UK by Fichtner Consulting Engineers Ltd 2004
14. Themelis, Nickolas J. An overview of the global waste-to-energy industry, Waste Management World 2003
15. from the homepage of the UK Renewable Energy Association
16. More recycling raises average energy content of waste used to generate electricity. U.S. Energy Information Administration. September 2012.
17. Directive 2009/28/EC on the promotion of the use of energy from renewable sources. European Union. April 23, 2009.
18. The biogenic content of process streams from mechanical–biological treatment plants producing solid recovered fuel. Do the manual sorting and selective dissolution determination methods correlate? by Mélanie Séverin, Costas A. Velis, Phil J. Longhurst and Simon J.T. Pollard., 2010. In: Waste Management 30(7): 1171-1182
19. A New Method to Determine the Ratio of Electricity Production from Fossil

and Biogenic Sources in Waste-to-Energy Plants. by Fellner, J., Cencic, O. and Rechberger, H., 2007. In: Environmental Science & Technology, 41(7): 2579-2586.

20. Determination of biogenic and fossil CO_2 emitted by waste incineration based on $^{14}CO_2$ and mass balances. by Mohn, J., Szidat, S., Fellner, J., Rechberger, H., Quartier, R., Buchmann, B. and Emmenegger, L., 2008. In: Bioresource Technology, 99: 6471-6479.
21. Fuelled stations and FMS (PDF). ofgem.gov.uk. Retrieved 28 February 2015.
22. Fuel Measurement and Sampling (FMS) Questionnaire: Carbon-14. ofgem.gov.uk. Retrieved 28 February 2015.
23. Energy from Waste State-of-the-Art Report, Statistics 5th Edition August 2006. International Solid Waste Association (ISWA)
24. Energy-from-Waste facility in Lee County run as Covanta Lee, Inc.
25. Algonquin Power Energy from Waste Facility from the homepage of Algonquin Power
26. Enerkem to Squeeze Biofuel Out of Old Electricity Poles. gigaom.com. January 13, 2009.
27. Enerkem press release December 2013 found at the PR Newswire webpage.
28. Enerkem Website. Enerkem. March 18, 2016.
29. AFSOC makes ‹green' history while investing in future. US Air Force Special Operations Command. Retrieved 2011-04-28..
30. Pyrogenesis Perfecting Plasma. Biomass Magazine.
31. http://www.govliquidation.com/auction/view?auctionId=6241220
32. http://www.rewmag.com/dod-gasification-auction-pyrogenesis.aspx
33. Autonomie energétique pour un refuge de montagne : panneaux solaires. Connaissance des Énergies. 5 July 2012. Retrieved 28 February 2015.
34. Refuge de Sarenne using wood-fired gasification boiler

Mechanical Biological Treatment

Chapter Contents

MBT stands for Mechanical Biological Treatment. It is a process that could help the UK to meet its obligations under the Landfill Directive (EC/ 31/1999), of diverting biodegradable municipal solid waste from landfill to help protect the environment and reduce the amount of greenhouse gas emitted from landfill sites.

MBT describes a number of different processes dealing with the biological treatment of waste. It is the combination of both biological and physical processes, which can be arranged in a number of different ways. MBT is an established waste treatment technology in many European countries such as Germany and Austria (Welsh Assembly Government).

INTRODUCTION

A **mechanical biological treatment** (MBT) system is a type of waste processing facility that combines a sorting facility with a form of biological treatment such as composting or anaerobic digestion. MBT plants are designed to process mixed household waste as well as commercial and industrial wastes.

PROCESS

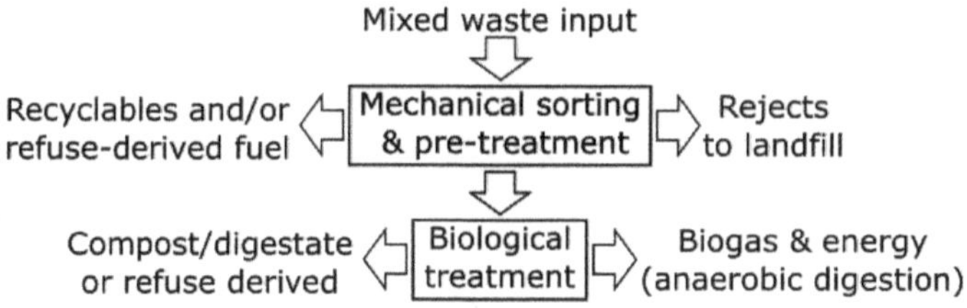

Process flow chart.

The terms *mechanical biological treatment* or *mechanical biological pre-treatment* relate to a group of solid waste treatment systems. These systems enable the recovery of materials contained within the mixed waste and facilitate the stabilisation of the biodegradable component of the material.

The sorting component of the plants typically resemble a materials recovery facility. This component is either configured to recover the individual elements of the waste or produce a Refuse-derived fuel that can be used for the generation of power.

The components of the mixed waste stream that can be recovered include:

- Ferrous Metal
- Non-ferrous metal
- Plastic
- Glass

TERMINOLOGY

MBT is also sometimes termed BMT – biological mechanical treatment – however this simply refers to the order of processing, i.e. the biological phase of the system precedes the mechanical sorting. MBT should not be confused with MHT – *mechanical heat treatment*

MECHANICAL SORTING

The "mechanical" element is usually an automated mechanical sorting stage. This either removes recyclable elements from a mixed waste stream (such as metals, plastics, glass and paper) or processes them. It typically involves factory style conveyors, industrial magnets, eddy current separators, trommels, shredders and other tailor made systems, or the sorting is done manually at hand picking stations. The mechanical element has a number of similarities to a materials recovery facility (MRF).

Some systems integrate a wet MRF to separate by density and floatation and to recover & wash the recyclable elements of the waste in a form that can be sent for recycling. MBT can alternatively process the waste to produce a high calorific fuel termed refuse derived fuel (RDF). RDF can be used in cement kilns or thermal combustion power plants and is generally made up from plastics and biodegradable organic waste. Systems which are configured to produce RDF include the Herhof and Ecodeco Processes. It is a common misconception that all MBT processes produce RDF. This is not the case and depends strictly on system configuration and suitable local markets for MBT outputs.

Wet material recovery facility, Hiriya, Israel.

BIOLOGICAL PROCESSING

Twin stage & UASB anaerobic digesters.

The "biological" element refers to either:

- Anaerobic digestion
- Composting
- Biodrying

Anaerobic digestion harnesses anaerobic microorganisms to break down the biodegradable component of the waste to produce biogas and soil improver. The biogas can be used to generate electricity and heat.

Biological can also refer to a composting stage. Here the organic component is broken down by naturally occurring aerobic microorganisms. They breakdown the waste into carbon dioxide and compost. There is no green energy produced by systems employing only composting treatment for the biodegradable waste.

In the case of biodrying, the waste material undergoes a period of rapid heating through the action of aerobic microbes. During this partial composting stage the heat generated by the microbes result in rapid drying of the waste. These systems are often configured to produce a refuse-derived fuel where a dry, light material is advantageous for later transport and combustion.

Some systems incorporate both anaerobic digestion and composting. This may either take the form of a full anaerobic digestion phase, followed by the maturation (composting) of the digestate. Alternatively a partial anaerobic digestion phase can be induced on water that is percolated through the raw waste, dissolving the readily available sugars, with the remaining material being sent to a windrow composting facility.

By processing the biodegradable waste either by anaerobic digestion or by composting MBT technologies help to reduce the contribution of greenhouse gases to global warming.

Usable wastes for this system:

- Municipal solid waste
- Commercial and industrial waste
- Sewage sludge

Possible products of this system:

- Renewable fuel (biogas) leading to renewable power
- Recovered recycable materials such as metals, paper, plastics, glass etc.
- Digestate - an organic fertiliser and soil improver
- Carbon credits – additional revenues
- High calorific fraction refuse derived fuel - Renewable fuel content dependent upon biological component
- Residual unusable materials prepared for their final safe treatment (e.g. incineration or gasification) and/or landfill

Further advantages:

- Small fraction of inert residual waste
- Reduction of the waste volume to be deposited to at least a half (density > 1.3 t/m^3), thus the lifetime of the landfill is at least twice as long as usually
- Utilisation of the leachate in the process
- Landfill gas not problematic as biological component of waste has been stabilised
- Daily covering of landfill not necessary

CONSIDERATION OF APPLICATIONS

MBT systems can form an integral part of a region's waste treatment infrastructure. These systems are typically integrated with kerbside collection schemes. In the event that a refuse-derived fuel is produced as a by-product then a combustion facility would be required. This could either be an incineration facility or a gasifier.

Alternatively MBT solutions can diminish the need for home separation and kerbside collection of recyclable elements of waste. This gives the ability of local authorities, municipalities and councils to reduce the use of waste vehicles on the roads and keep recycling rates high.

POSITION OF ENVIRONMENTAL GROUPS

Friends of the Earth suggests that the best environmental route for residual waste is to firstly maximise removal of remaining recyclable materials from the waste stream (such as metals, plastics and paper). The amount of waste remaining should be composted or anaerobically digested and disposed of to landfill, unless sufficiently clean to be used as compost.

A report by Eunomia undertook a detailed analysis of the climate impacts of different residual waste technologies. It found that an MBT process that extracts both the metals and plastics prior to landfilling is one of the best options for dealing with our residual waste, and has a lower impact than either MBT processes producing RDF for incineration or incineration of waste without MBT.

Friends of the Earth does not support MBT plants that produce refuse derived fuel (RDF), and believes MBT processes should occur in small, localised treatment plants.

MUNICIPAL SOLID WASTE

Municipal solid waste (**MSW**), commonly known as **trash** or **garbage** in the United States and as **refuse** or **rubbish** in Britain, is a waste type consisting of everyday items that are discarded by the public. "Garbage" can also refer specifically to food waste, as in a garbage disposal; the two are sometimes collected separately.

Composition

The composition of municipal solid waste varies greatly from municipality to municipality and changes significantly with time. In municipalities which have a well developed waste recycling system, the waste stream consists mainly of intractable wastes such as plastic film, and non-recyclable packaging materials. At the start of the 20th century, the majority of domestic waste (53%) in the UK consisted of coal ash from open fires. In developed areas without significant recycling activity it predominantly includes food wastes, market wastes, yard wastes, plastic containers and product packaging materials, and other miscellaneous solid wastes from residential, commercial, institutional, and industrial sources. Most definitions of municipal solid waste do not include industrial wastes, agricultural wastes, medical waste, radioactive waste or sewage sludge. Waste collection is performed by the municipality within a given area. The term *residual waste* relates to waste left from household sources containing materials that have not been separated out or sent for reprocessing. Waste can be classified in several ways but the following list represents a typical classification:

- Biodegradable waste: food and kitchen waste, green waste, paper (most can be recycled although some difficult to compost plant material may be excluded)
- Recyclable materials: paper, cardboard, glass, bottles, jars, tin cans, aluminum cans, aluminum foil, metals, certain plastics, fabrics, clothes, tires, batteries, etc.
- Inert waste: construction and demolition waste, dirt, rocks, debris
- Electrical and electronic waste (WEEE) - electrical appliances, light bulbs, washing machines, TVs, computers, screens, mobile phones, alarm clocks, watches, etc.
- Composite wastes: waste clothing, Tetra Packs, waste plastics such as toys
- Hazardous waste including most paints, chemicals, tires, batteries, light bulbs, electrical appliances, fluorescent lamps, aerosol spray cans, and fertilizers
- Toxic waste including pesticides, herbicides, and fungicides
- Biomedical waste, expired pharmaceutical drugs, etc.

Components of Solid Waste Management

Bins to collect paper, aluminium, glass, PET bottles and incinerable waste.

The municipal solid waste industry has four components: recycling, composting, disposal, and waste-to-energy via incineration. There is no single approach that can be applied to the management of all waste streams, therefore the Environmental Protection Agency, federal agency of the United States of America, developed a hierarchy ranking strategy for municipal solid waste. The Waste Management Hierarchy is made up of four levels ordered from most preferred to least preferred methods based on their environmental soundness: Source reduction and reuse; recycling or composting; energy recovery; treatment and disposal.

Collection

The functional element of collection includes not only the gathering of solid waste and recyclable materials, but also the transport of these materials, after collection, to the location where the collection vehicle is emptied. This location may be a materials processing facility, a transfer station or a landfill disposal site.

Waste handling and Separation, Storage and Processing at the Source

Waste handling and separation involves activities associated with waste management until the waste is placed in storage containers for collection. Handling also encompasses the movement of loaded containers to the point of collection. Separating different types of waste components is an important step in the handling and storage of solid waste at the source.

Segregation and Processing and Transformation of solid Wastes

The types of means and facilities that are now used for the recovery of waste materials that have been separated at the source include curbside ('kerbside' in the UK) collection, drop-off and buy-back centers. The separation and processing of wastes that have been separated at the source and the separation of commingled wastes usually occur at a materials recovery facility, transfer stations, combustion facilities and disposal sites.

Transfer and Transport

This element involves two main steps. First, the waste is transferred from a smaller collection vehicle to larger transport equipment. The waste is then transported, usually over long distances, to a processing or disposal site.

Disposal

Today, the disposal of wastes by land filling or land spreading is the ultimate fate of all solid wastes, whether they are residential wastes collected and transported directly to a landfill site, residual materials from materials recovery facilities (MRFs), residue from the combustion of solid waste, compost, or other substances from various solid waste processing facilities. A modern sanitary landfill is not a dump; it is an engineered facility used for disposing of solid wastes on land without creating nuisances or hazards to public health or safety, such as the problems of insects and the contamination of ground water.

Mixed municipal waste, Hiriya, Tel Aviv.

Reusing

In the recent years environmental organizations, such as Freegle or Freecycle Network, have been gaining popularity for their online reuse networks. These networks provide a worldwide online registry of unwanted items that would otherwise be thrown away, for individuals and nonprofits to reuse or recycle. Therefore, this free Internet-based service reduces landfill pollution and promotes the gift economy.

Landfills

Landfills are created by land dumping. Land dumping methods vary, most commonly it involves the mass dumping of waste into a designated area, usually a hole or sidehill. After the waste is dumped, it is then compacted by large machines. When the dumping cell is full, it is then "sealed" with a plastic sheet and covered in several feet of dirt. This is the primary method of dumping in the United States because of the low cost and abundance of unused land in North America. Landfills pose the threat of pollution, and can intoxicate ground water. The signs of pollution are effectively masked by disposal companies and it is often hard to see any evidence. Usually landfills are surrounded by large walls or fences hiding the mounds of debris. Large amounts of chemical odor eliminating agent are sprayed in the air surrounding landfills to hide the evidence of the rotting waste inside the plant.

Energy Generation

Municipal solid waste can be used to generate energy. Several technologies have been developed that make the processing of MSW for energy generation cleaner and more economical than ever before, including landfill gas capture, combustion, pyrolysis, gasification, and plasma arc *gasification*. While older waste incineration plants emitted a lot of pollutants, recent regulatory changes and new technologies have significantly reduced this concern. United States Environmental Protection Agency (EPA) regulations in 1995 and 2000 under the Clean Air Act have succeeded in reducing emissions of dioxins from waste-to-energy facilities by more than 99 percent below 1990 levels, while mercury emissions have been reduced by over 90 percent. The EPA noted these improvements in 2003, citing waste-to-energy as a power source "with less environmental impact than almost any other source of electricity".

LANDFILL

A landfill in Poland.

A **landfill site** (also known as a **tip**, **dump**, **rubbish dump**, **garbage dump** or **dumping ground** and historically as a **midden**) is a site for the disposal of waste materials by burial and is the oldest form of waste treatment (although the burial part is modern; historically, refuse was just left in piles or thrown into pits). Historically, landfills have been the most common method of organized waste disposal and remain so in many places around the world.

Some landfills are also used for waste management purposes, such as the temporary storage, consolidation and transfer, or processing of waste material (sorting, treatment, or recycling).

A landfill also may refer to ground that has been filled in with rocks instead of waste materials, so that it can be used for a specific purpose, such as for building houses. Unless they are stabilized, these areas may experience severe shaking or soil liquefaction of the ground during a large earthquake.

Operations

One of several landfills used by Dryden, Ontario, Canada.

Typically, operators of well-run landfills for non-hazardous waste meet predefined specifications by applying techniques to:

- confine waste to as small an area as possible
- compact waste to reduce volume
- cover waste (usually daily) with layers of soil

During landfill operations a scale or weighbridge may weigh waste-collection vehicles on arrival and personnel may inspect loads for wastes that do not accord with the landfill's waste-acceptance criteria. Afterward, the waste-collection vehicles use the existing road network on their way to the tipping face or working front, where they unload their contents. After loads are deposited, compactors or bulldozers can spread and compact the waste on the working face. Before leaving the landfill boundaries, the waste collection vehicles may pass through a wheel-cleaning facility. If necessary, they return to the weighbridge for re-weighing without their load. The weighing process can assemble statistics on the daily incoming waste-tonnage, which databases can retain for record keeping. In addition to trucks, some landfills may have equipment to handle railroad containers. The use of "rail-haul" permits landfills to be located at more remote sites, without the problems associated with many truck trips.

Typically, in the working face, the compacted waste is covered with soil or alternative materials daily. Alternative waste-cover materials include chipped wood or other "green waste", several sprayed-on foam products, chemically "fixed" bio-solids, and temporary blankets. Blankets can be lifted into place at night and then removed the following day prior to waste placement. The space that is occupied daily by the compacted waste and the cover material is called a daily cell. Waste compaction is critical to extending the life of the landfill. Factors such as waste compressibility, waste-layer thickness and the number of passes of the compactor over the waste affect the waste densities.

Advantages

Landfills are often the most cost-efficient way to dispose of waste, especially in countries like the United States with large open spaces. While resource recovery and incineration both require extensive investments in infrastructure, and material recovery also requires extensive manpower to maintain, landfills have fewer fixed—or ongoing—costs, allowing them to compete favorably. In addition, landfill gas can be upgraded to natural gas—landfill gas utilization—which is a potential revenue stream. Another advantage is having a specific location for disposal that can be monitored, where waste can be processed to remove all recyclable materials before tipping.

Social and Environmental Impact

Landfill operation in Hawaii. Note that the area being filled is a single, well-defined "cell" and that a protective landfill liner is in place (exposed on the left) to prevent contamination by leachates migrating downward through the underlying geological formation.

Landfills have the potential to cause a number of issues. Infrastructure disruption, such as damage to access roads by heavy vehicles, may occur. Pollution of local roads and water courses from wheels on vehicles when they leave the landfill can be significant and can be mitigated by wheel washing systems. Pollution of the local environment, such as contamination of groundwater or aquifers or soil contamination may occur, as well.

Leachate

Extensive efforts are made to capture and treat leachate from landfills before it reaches groundwater aquifers, but engineered liners always have a lifespan, though it may be 100 years or more. Eventually, every landfill liner will leak,allowing the leachate to contaminate the groundwater. Installation of composite liners with flexible membrane and soil barrier is enforced by the EPA to ensure that leachate is withheld.

Dangerous gases

Rotting food and other decaying organic waste allows methane and carbon dioxide to seep out of the ground and up into the air. Methane is a potent greenhouse gas, and can itself be a danger because it is flammable and potentially explosive. In properly managed landfills, gas is collected and utilized. This could range from simple flaring to landfill gas utilization. Carbon dioxide is the most widely produced greenhouse gas. It traps heat in the atmosphere, contributing to climate change.

Infections

Poorly run landfills may become nuisances because of vectors such as rats and flies which can cause infectious diseases. The occurrence of such vectors can be mitigated through the use of daily cover.

Other potential issues include wildlife disruption, dust, odor, noise pollution, and reduced local property values.

Landfill gas

Gases are produced in landfills due to the anaerobic digestion by microbes. In a properly managed landfill this gas is collected and used. Its uses range from simple flaring to the landfill gas utilization and generation of electricity. Landfill gas monitoring alerts workers to the presence of a build-up of gases to a harmful level. In some countries, landfill gas recovery is extensive; in the United States, for example, more than 850 landfills have active landfill gas recovery systems.

Regional Practice

A landfill in Perth, Western Australia.

South East New Territories Landfill, Hong Kong

Canada

Landfills in Canada are regulated by provincial environmental agencies and environmental protection acts (EPA). Older facilities tend to fall under current standards and are monitored for leaching. Some former locations have been converted to parkland.

European Union

In the European Union, individual states are obliged to enact legislation to comply with the requirements and obligations of the European Landfill Directive. In the UK this is the Waste Implementation Programme.

United Kingdom

Landfilling practices in the UK have had to change in recent years to meet the challenges of the European Landfill Directive. The UK now imposes landfill tax upon biodegradable waste which is put into landfills. In addition to this the Landfill Allowance Trading Scheme has been established for local authorities to trade landfill quotas in England. A different system operates in Wales where authorities are not able to 'trade' between themselves, but have allowances known as the Landfill Allowance Scheme.

United States

U.S. landfills are regulated by each state's environmental agency, which establishes minimum guidelines; however, none of these standards may fall below those set by the United States Environmental Protection Agency (EPA).

Permitting a landfill generally takes between 5 and 7 years, costs millions of dollars and requires rigorous siting, engineering and environmental studies and demonstrations to ensure local environmental and safety concerns are satisfied.

Microbial Topics

The status of a landfill's microbial community may determine its digestive efficiency.

Bacteria that digest plastic have been found in landfills.

Reclaiming Materials

Landfills can be regarded as a viable and abundant source of materials and energy. In the developing world, waste pickers often scavenge for still-usable materials. In a commercial context, landfill sites have also been discovered by companies, and many have begun harvesting materials and energy . Well known examples are gas recovery facilities. Other commercial facilities include waste incinerators which have built-in material recovery. This material recovery is possible through the use of filters (electro filter, active carbon and potassium filter, quench, HCl-washer, SO_{2-}washer, bottom ash-grating, etc.).

Alternatives

In addition to waste reduction and recycling strategies, there are various alternatives to landfills, including waste-to-energy incineration, anaerobic digestion, composting, mechanical biological treatment, pyrolysis and plasma arc gasification, which have all begun to establish themselves in the market. Depending on local economics and incentives, these can be made more financially attractive than landfills.

Restrictions

Countries including Germany, Austria, Sweden, Denmark, Belgium, the Netherlands, and Switzerland, have banned the disposal of untreated waste in landfills. In these countries, only the ashes from incineration or the stabilized output of mechanical biological treatment plants may still be deposited.

REFERENCES

1. Centenary history of waste and waste managers in London and south east England- Page 7
2. Non-hazardous Waste U.S. Environmental Protection Agency, Municipal Solid Waste
3. Municipal Solid Waste U.S. Energy Information Administration
4. Mechanical Biological Treatment Welsh Assembly (2005) Mechanical Biological Treatment, Environment Countryside and Planning Website, Welsh Assembly
5. Organics -Green Bin . Christchurch City Council. Retrieved 19 March 2016.
6. Nonhazardous waste U.S. Energy Information Administration
7. Advancing Sustainable Materials Management: Facts and Figures
8. Rogers, Heather. Gone Tomorrow: The Hidden Life of Garbage.
9. Environmental and Energy Study Institute Issue Brief
10. Combustion Emissions from Hazardous Waste Incinerators, Boilers and Industrial Furnaces, and Municipal Solid Waste Incinerators- Results from Five STAR Grants and Research Needs U.S. EPA
11. U.S. EPA Letter to Maria Zannes, President, Integrated Waste Services Association

12. Midden . Merriam-Webster. Retrieved 18 May 2014.
13. Alternative Daily Cover (ADC) . Retrieved September 14, 2012.
14. LFG Energy Projects . United States Environmental Protection Agency. Retrieved February 21, 2015.
15. US EPA, "Solid Waste Disposal Facility Criteria; Proposed Rule", Federal Register 53(168):33314–33422, 40 CFR Parts 257 and 258, US EPA, Washington, D.C., August 30 (1988a).
16. Landfill Leachate Corrodes Government Structures And Cities
17. CO2 101: Why is carbon dioxide bad? . Mother Nature Network. Retrieved 2016-11-30.
18. Powell, Jon T.; Townsend, Timothy G.; Zimmerman, Julie B. (21 September 2015). Estimates of solid waste disposal rates and reduction targets for landfill gas emissions . Nature Climate Change. 6: 162–165. doi:10.1038/nclimate2804.
19. Landfill Inventory Management Ontario - How Ontario regulates Landfills - Ministry of the Environment
20. Aging Landfills: Ontario's Forgotten Polluters - Eco Issues
21. Modern landfills . Retrieved February 21, 2015.
22. Gomez, A.M.; =Yannarell, A.C.; Sims, G.K.; Cadavid-Resterpoa, G.; Herrera, C.X.M. (2011). Characterization of bacterial diversity at different depths in the Moravia Hill Landfill site at Medellín, Colombia . Soil Biology and Biochemistry. 43 (6): 1275–1284. doi:10.1016/j.soilbio.2011.02.018.
23. Gwyneth Dickey Zaikab (March 2011). Marine microbes digest plastic .
24. Multiple Purpose industries using landfills for energy
25. Commercial exploitation of gas from landfills

Index